CHEMICAL ENGINEERING METHODS AND TECHNOLOGY

HEAT CONDUCTION

METHODS, APPLICATIONS AND RESEARCH

Chemical Engineering Methods and Technology

Additional books and e-books in this series can be found on Nova's website under the Series tab.

CHEMICAL ENGINEERING METHODS AND TECHNOLOGY

HEAT CONDUCTION

METHODS, APPLICATIONS AND RESEARCH

JORDAN HRISTOV

AND

RACHID BENNACER

EDITORS

nova science publishers

New York

Additional color graphics may be available in the e-book version of this book.

Library of Congress Cataloging-in-Publication Data

Names: Hristov, Jordan, editor. | Bennacer, Rachid, editor.
Title: Heat conduction: methods, applications and research / editors, Jordan Hristov, Ph.D., D.Sc., Department of Chemical Engineering, University of Chemical Technology and Metallurgy, Sofia, Bulgaria and Rachid Bennacer, University of Paris-Saclay, Orsay, France.
Description: Hauppauge, New York, USA: Nova Science Publishers, Inc., [2019]
| Series: Chemical engineering methods and technology | Includes bibliographical references and index.
Identifiers: LCCN 2018050907 (print) | LCCN 2018053628 (ebook) | ISBN 9781536146745 (ebook) | ISBN 9781536146738 | ISBN 9781536146745 (ebook)
Subjects: LCSH: Heat--Conduction.
Classification: LCC QC321 (ebook) | LCC QC321 .H43 2019 (print) | DDC 536/.23--dc23
LC record available at https://lccn.loc.gov/2018050907

Published by Nova Science Publishers, Inc. † New York

CONTENTS

PREFACE

Heat conduction problems are classic challenges for engineers and scientist with consciously emerging challenging problems to be solved. Among the multifaceted applications of heat conduction problem there principle, say classical, problems that are always challenging to develop new solutions, among them: transient heat conductions with non-linear heat conductivities, Stefan problems, phase change materials, porous media heat conduction with convection, etc.

This collection of studies arranged as a book was suggested by the publisher and when we, the editors, were invited there was an ambiguity what exactly to be included as topic. Then, we decided to invite authors who are appearing in journal publications and the responses as chapters arranged in this book clearly indicate what some of the modern trend in conduction studies are.

Following these introductory notes we may mention three chapters applying the integral-balance method, which is one of the most powerful engineering methods in heat conduction.

In Chapter 1, Bolati et al. applied different approximations for the one-dimensional one-phase Stefan problem corresponding to the fusion process of a semi-infinite material with a temperature boundary condition at the fixed face and non-linear temperature-dependent thermal conductivity. A refined balance integral method, assuming a quadratic temperature profile in space is successfully applied.

Agrawal and Sahu, in Chapter 2, applied the variational integral method to analyze variety of rewetting problems. The analysis is extended to include the effect of precursory cooling, variation in heat transfer in multiple step functions, exponential functions and variation in property. Based on the analysis, closed form solution for the temperature field and rewetting velocity are developed.

In Chapter 3, Hristov applied a constitutive approach to modeling of heat shock waves described by the diffusion approximation in radiation heat transfer in terms of time fractional derivatives. Approximate closed-form analytical solutions by a double integration method concerning a step change of the surface temperature and two classical problems with time-dependent boundary are developed for special case of time-depended boundary conditions.

Avci et al. (Chapter 4), considered the heat conduction equation in terms of Caputo-Fabrizio operator with a nonsingular exponential kernel in problems formulated in a line segment. The solutions developed by the Laplace transform in time and the finite sin-Fourier transform in space allowed demonstrating the effect of the fractional parameters controlling the memory kernel, thus, elucidating what really as physics the Caputo-Fabrizio operator models in heat conduction.

In Chapter 5, Raguii et al. summarize some recent results on the conduction-limit impact within a double-diffusive convection of a Bingham plastic fluid which occurs within a Silicon Carbide Ceramic annulus; contained between a cold "and less concentric" outer circular cylinder and a hot "and concentric" inner one. Predictions of a such critical limit, into these few pages, will be a function of some relevant parameters; such as the boundary layers' ratio, known generally as Lewis number and the thermosolutal buoyancy one; which is a pure relation between the thermosolutal gradients and its expansion coefficients, as long as the geometry aspect ratio is developed.

Abahri and Ma consider in Chapter 6 the current state of knowledge related to transport of heat, fluid (air) and mass (moisture) in porous construction materials.

Finally, Ma et al. (Chapter 7) consider the energy storage properties of Phase Change Materials (PCMs) related to energy consumption in buildings. Analytical solution in semi-infinity domain, a short time solution in 2D rectangular corner, a semi-analytical solution for a bounded domain, the multi-phase problem in a Cartesian domain, and the phase change problem in a cylindrical domain are developed.

After these brief outlines of the main results presented in the chapters collected we may express out warm appreciations to all author contributing the book. Moreover, last but not least, we like to express our gratitude to the publishing office for their understanding about the problems emerging during the manuscript collection and editing. We believe that this book would a good reference source to many researchers involved in various s application s of heat conduction problems and would motivate them to go further in this amazing and challenging area of research.

Prof. Dr. Jordan Hristov, PhD, DSc
Department of Chemical Engineering,
University of Chemical Technology and Metallurgy,
Sofia, Bulgaria

Prof. Dr. Rachid Bennacer, PhD
LMT-ENS Cachan, CNRS, Paris-Saclay University,
Cachan, France

July 2018

In: Heat Conduction
Editors: J. Hristov and R. Bennacer
ISBN: 978-1-53614-673-8

Chapter 1

APPROXIMATE SOLUTIONS TO THE ONE-PHASE STEFAN PROBLEM WITH NON-LINEAR TEMPERATURE-DEPENDENT THERMAL CONDUCTIVITY

***Julieta Bollati*[1,2,*], *María F. Natale*[2], *José A. Semitiel*[2] and *Domingo A. Tarzia*[1,2,†]**
[1]Consejo Nacional de Investigaciones Científicas y Tecnológicas
[2]Departamento de Matemática, Universidad Austral,
Rosario, Argentina

Abstract

In this chapter we consider different approximations for the one-dimensional one-phase Stefan problem corresponding to the fusion process of a semi-infinite material with a temperature boundary condition at the fixed face and non-linear temperature-dependent thermal conductivity. The knowledge of the exact solution of this problem, allows to compare it directly with the approximate solutions obtained by applying the heat balance integral method, an alternative form to it and the refined balance integral method, assuming a quadratic temperature profile in space. In

*Corresponding Author Email: JBollati@austral.edu.ar.
†E-mail address: DTarzia@austral.edu.ar.

all cases, the analysis is carried out in a dimensionless way by the Stefan number (Ste) parameter.

Keywords: Stefan problem, heat balance integral method, refined integral method, temperature-dependent thermal conductivity, exact solutions

1. Introduction

Stefan problems model heat transfer processes that involve a change of phase. They constitute a broad field of study since they arise in a great number of mathematical and industrial significance problems. Phase-change problems appear frequently in industrial processes and other problems of technological interest [1]-[4]. Reviews on the subject were given in [5, 6].

The heat balance integral method introduced in [7] is a well-known method of approximation of solutions of Stefan problems. It transforms the heat equation into an ordinary differential equation over time assuming a quadratic temperature profile in space. For these temperature profiles, different variants of this method were established by [8]. Moreover in [9]-[14] the heat balance integral method is used for different temperature profiles.

In this chapter, we obtain approximate solutions to a phase-change Stefan problem (2)-(6) for a non-linear heat conduction equation corresponding to a semi-infinite region $x > 0$ with a thermal conductivity $k(\theta)$ given by

$$k(\theta) = \frac{\rho c}{(a + b\theta)^2} \tag{1}$$

and phase change temperature $\theta_f = 0$. This kind of thermal conductivity or diffusion coefficient was considered in [15]-[25].

The exact solution of this problem was given in [26], where the temperature is the unique solution of an integral equation and the coefficient that characterizes the free boundary is the unique solution of a transcendental equation. From this fact, the most remarkable aspect of this chapter lies in the comparison of the exact solution, which is difficult and cumbersome to operate, with different approaches obtained through: the heat balance integral method, an alternative form to it [8] and the refined integral method [27].

The methods mentioned above have been developed for the non-linear diffusion equation to the case of a linearly temperature-dependent thermal diffusivity [28] and a power-law dependent diffusivity with integer positive exponent [29], [30]; obtaining closed forms of approximate solutions.

The goal of this chapter is to provide approximate solutions in order to facilitate the search of the solution and show that it is worth using approximate methods due to the small error with respect to the exact solution. In all the applied methods the dimensionless parameter called Stefan number is defined. We take Stefan number up to 1 due to the fact that it covers most of phase change materials [31].

2. Mathematical Formulation and Exact Solution

We consider a one-dimensional one-phase Stefan problem for the fusion of a semi-infinite material $x > 0$ with non-linear temperature-dependent thermal conductivity. This problem can be formulated mathematically in the following way:

Problem (P). Find the temperature $\theta = \theta(x,t)$ at the liquid region $0 < x < s(t)$ and the evolution of the moving separation phase given by $x = s(t)$ satisfying the following conditions

$$\rho c \frac{\partial \theta}{\partial t} = \frac{\partial}{\partial x}\left(k(\theta)\frac{\partial \theta}{\partial x}\right), \qquad 0 < x < s(t), \quad t > 0 \tag{2}$$

$$\theta = \theta_0 > 0 \ , \qquad \text{on } x = 0, \quad t > 0 \tag{3}$$

$$k\left(\theta\right)\frac{\partial \theta}{\partial x} = -\rho\lambda\dot{s}(t) \ , \qquad \text{on } x = s(t), \quad t > 0 \tag{4}$$

$$\theta = 0 \ , \qquad \text{on } x = s(t), \quad t > 0 \tag{5}$$

$$s(0) = 0 \tag{6}$$

where the parameters c, ρ and λ are the specific heat, the density and the latent heat of fusion of the medium respectively, all of them assumed to be positive constants. The thermal conductivity k is given by (1), with positive parameters a and b.

In [26] was proved the existence and uniqueness of an exact solution of the similarity type of the free boundary problem $(2) - (6)$ for $t \geq t_0 > 0$ with t_0 an arbitrary positive time when data satisfy condition $ac = b\lambda$.

If we define the non-dimensional Stefan number by

$$\text{Ste} = \frac{c\theta_0}{\lambda}, \tag{7}$$

then we have $\text{Ste} = \frac{b\theta_0}{a}$.

Now, we can write the exact solution as [26]:

$$\theta(x,t) = \frac{1}{b}\left[\frac{1}{\Theta(x,t)} - a\right] \ , \ 0 < x < s(t) \ , \ t > 0, \tag{8}$$

$$s(t) = \frac{2}{a}\xi\sqrt{t} \ , \ t > 0, \tag{9}$$

where Θ is the unique solution in variable x of the following integral equation

$$\Theta(x,t) = \frac{1}{a}\left(1 + \frac{\text{Ste}}{(1+\text{Ste})\text{erf}(\Lambda)}\text{erf}\left(\frac{\int_0^x \frac{d\eta}{\Theta(\eta,t)}}{2\sqrt{t}} - \Lambda\right)\right) \ , \ 0 \le x \le s(t), \tag{10}$$

for $t \ge t_0 > 0$ with t_0 an arbitrary positive time and ξ is given by

$$\xi = \frac{2\Lambda \exp\left(\Lambda^2\right)}{1 + \text{Ste}}, \tag{11}$$

where Λ is the unique positive solution to the following equation

$$z\exp(z^2)\text{erf}(z) = \frac{\text{Ste}}{\sqrt{\pi}} \quad , \quad z > 0. \tag{12}$$

Remark 2.1. *In [26] was proved that the integral equation (10) is equivalent to solve the following Cauchy differential problem in variable x:*

$$\begin{cases} \dfrac{\partial Y}{\partial x}(x,t) = \dfrac{a}{2\sqrt{t}\left[1 + \frac{Ste}{(1+Ste)erf(\Lambda)} erf(Y(x,t))\right]}, & 0 < x < s(t), \ t > 0, \\ Y(0,t) = -\Lambda, \end{cases} \tag{13}$$

where

$$Y(x,t) = \frac{\int_0^x \frac{d\eta}{\Theta(\eta,t)}}{2\sqrt{t}} - \Lambda \tag{14}$$

with a positive parameter $t \ge t_0 > 0$ and Λ the unique solution of (12).

3. Heat Balance Integral Methods

As one of the mechanisms for the heat conduction is the diffusion, the excitation at the fixed face $x = 0$ (for example, a temperature, a flux or a convective condition) does not spread instantaneously to the material $x > 0$. However, the effect of the fixed boundary condition can be perceived in a bounded interval $[0, \delta(t)]$ (for every time $t > 0$) outside of which the temperature remains equal to the initial temperature. The heat balance integral method presented in [7] established the existence of a function $\delta = \delta(t)$ that measures the depth of the thermal layer. In problems with a phase of change, this layer is assumed as the free boundary, i.e., $\delta(t) = s(t)$.

From equation (2) and conditions (4) and (5) we obtain the new condition:

$$k(\theta)\left(\frac{\partial \theta}{\partial x}\right)^2 = \frac{\lambda}{c}\frac{\partial}{\partial x}\left(k(\theta)\frac{\partial \theta}{\partial x}\right) \qquad \text{on} \quad x = s(t), \ t > 0. \tag{15}$$

From equation (2) and conditions (3), (4) and (5) we obtain the integral condition:

$$\begin{aligned} \frac{d}{dt}\int_0^{s(t)} \theta(x,t)dx &= \int_0^{s(t)} \frac{\partial \theta}{\partial t}(x,t)dx + \theta(s(t),t)\dot{s}(t) \\ &= \frac{1}{\rho c}\int_0^{s(t)} \frac{\partial}{\partial x}\left(k(\theta(x,t))\frac{\partial \theta}{\partial x}(x,t)\right)dx \\ &= \frac{-1}{\rho c}\left[\rho\lambda\dot{s}(t) + k(\theta_0)\frac{\partial \theta}{\partial x}(0,t)\right]. \end{aligned} \tag{16}$$

The classical heat balance integral method introduced in [7] proposes to approximate problem (P) through the resolution of a problem that arises on replacing the equation (2) by the equation (16), and the condition (4) by the condition (15); that is, the resolution of the approximate problem defined as follows: conditions (3), (5), (6), (15) and (16).

In [8], a variant of the classical heat balance integral method was proposed by replacing equation (2) by condition (16), keeping all others conditions of the problem (P) equals; that is, the resolution of an approximate problem defined as follows: conditions (3),(4),(5),(6) and (16).

From equation (2) and conditions (3) and (5) we also obtain:

$$\begin{aligned}
\int_0^{s(t)}\int_0^{x} \frac{\partial \theta}{\partial t}(\eta,t)d\eta dx &= \int_0^{s(t)}\int_0^{x} \frac{1}{\rho c}\frac{\partial}{\partial \eta}\left(k\left(\theta(\eta,t)\right)\frac{\partial \theta}{\partial \eta}(\eta,t)\right)d\eta dx \\
&= \int_0^{s(t)} \frac{1}{\rho c}\left[k\left(\theta(x,t)\right)\frac{\partial \theta}{\partial x}(x,t) - k\left(\theta_0\right)\frac{\partial \theta}{\partial x}(0,t)\right]dx \\
&= \frac{1}{\rho c}\int_0^{s(t)} \rho c\frac{\frac{\partial \theta}{\partial x}(x,t)}{(a+b\theta(x,t))^2}dx - \frac{k\left(\theta_0\right)}{\rho c}\frac{\partial \theta}{\partial x}(0,t)s(t) \\
&= -\frac{\theta_0\left(1+\text{Ste}\right)+\frac{\partial \theta}{\partial x}(0,t)s(t)}{a^2\left(1+\text{Ste}\right)^2}. \qquad (17)
\end{aligned}$$

The refined integral method introduced in [27] proposes to approximate problem (P) through the resolution of the approximate problem that arises by replacing equation (2) by (17), keeping all others conditions of the problem (P) equals. It is defined as follows: conditions (3), (4), (5), (6) and (17).

For solving the approximate problems previously defined we propose a quadratic temperature profile in space as follows:

$$\widetilde{\theta}(x,t) = \widetilde{A}\theta_0\left(1-\frac{x}{\widetilde{s}(t)}\right) + \widetilde{B}\theta_0\left(1-\frac{x}{\widetilde{s}(t)}\right)^2, \qquad (18)$$

where $\widetilde{\theta}$ and $\widetilde{s}$ are approximations of θ and s respectively.

Taking advantage of the fact of having the exact temperature of the problem (P), it is physically reasonable to impose that the approximate temperature given by (18) behaves in a similar manner than the exact one given by (8); that is: its sign, monotony and convexity in space. As θ verifies the following properties:

$$\theta(x,t) > 0, \qquad (19)$$

$$\frac{\partial \theta}{\partial x}(x,t) = -\frac{\theta_0}{a\text{Ste}}\frac{1}{\Theta^2(x,t)}\frac{\partial \Theta}{\partial x}(x,t) < 0, \qquad (20)$$

$$\frac{\partial^2 \theta}{\partial x^2}(x,t) = -\frac{\theta_0}{a\text{Ste}}\left(-\frac{2}{\Theta^3(x,t)}\frac{\partial \Theta}{\partial x}(x,t) + \frac{1}{\Theta^2(x,t)}\frac{\partial^2 \Theta}{\partial x^2}(x,t)\right) > 0, \qquad (21)$$

on $0 < x < s(t), t > 0$, we enforce the following conditions on $\widetilde{\theta}$:

$$\widetilde{\theta}(x,t) > 0, \tag{22}$$

$$\frac{\partial \widetilde{\theta}}{\partial x}(x,t) = -\frac{\theta_0}{\widetilde{s}(t)}\left(\widetilde{A} + 2\widetilde{B}\left(1 - \frac{x}{\widetilde{s}(t)}\right)\right) < 0, \tag{23}$$

$$\frac{\partial^2 \widetilde{\theta}}{\partial x^2}(x,t) = \frac{2\widetilde{B}\theta_0}{\widetilde{s}^2(t)} > 0, \tag{24}$$

for all $0 < x < \widetilde{s}(t), t > 0$. Therefore, we obtain that both constants $\widetilde{A}$ and $\widetilde{B}$ must be positive.

3.1. Approximate Solution Using the Classical Heat Balance Integral Method

The classical heat balance integral method proposes to approximate problem (P) through the resolution of the approximate problem defined in Sect. 3, that is:

Problem (P1). Find the temperature $\theta_1 = \theta_1(x,t)$ at the liquid region $0 < x < s_1(t)$ and the location of the free boundary $x = s_1(t)$ such that:

$$\frac{d}{dt}\int_0^{s_1(t)} \theta_1(x,t)dx = \frac{-1}{\rho c}\left[\rho\lambda\dot{s}_1(t) + k\left(\theta_0\right)\frac{\partial \theta_1}{\partial x}(0,t)\right], \quad 0 < x < s_1(t), \tag{25}$$

$$\theta_1 = \theta_0, \qquad \text{on } x = 0, \tag{26}$$

$$k\left(\theta_1\right)\left(\frac{\partial \theta_1}{\partial x}\right)^2 = \frac{\lambda}{c}\frac{\partial}{\partial x}\left(k(\theta_1)\frac{\partial \theta_1}{\partial x}\right), \qquad \text{on } x = s_1(t), \tag{27}$$

$$\theta_1 = 0, \qquad \text{on } x = s_1(t), \tag{28}$$

$$s_1(0) = 0. \tag{29}$$

By proposing the following quadratic temperature profile in space:

$$\theta_1(x,t) = \theta_0 A_1\left(1 - \frac{x}{s_1(t)}\right) + \theta_0 B_1\left(1 - \frac{x}{s_1(t)}\right)^2, \quad 0 < x < s_1(t),\ t > 0, \tag{30}$$

the free boundary is obtained of the form:

$$s_1(t) = \frac{2}{a}\xi_1\sqrt{t}, \quad t > 0, \tag{31}$$

where the constants A_1, B_1 y ξ_1 will be determined from the conditions (25), (26) and (27) of the problem (P1). Because of (30) and (31), the conditions (28) and (29) are immediately satisfied. From conditions (25) and (26) we obtain:

$$A_1 = \frac{2\left[3\text{Ste} - (1+\text{Ste})^2\xi_1^2(\text{Ste}+3)\right]}{\text{Ste}\left[(1+\text{Ste})^2\xi_1^2+3\right]}, \tag{32}$$

$$B_1 = \frac{3\left[-\text{Ste} + (1+\text{Ste})^2\xi_1^2(\text{Ste}+2)\right]}{\text{Ste}\left[(1+\text{Ste})^2\xi_1^2+3\right]}. \tag{33}$$

From the fact that $A_1 > 0$ and $B_1 > 0$ we obtain that $0 < \xi_1 < \xi^{\max}$ and $\xi_1 > \xi^{\min} > 0$, respectively where:

$$\xi^{\min} = \sqrt{\frac{\text{Ste}}{(1+\text{Ste})^2(2+\text{Ste})}}, \qquad \xi^{\max} = \sqrt{\frac{3\text{Ste}}{(1+\text{Ste})^2(3+\text{Ste})}}. \tag{34}$$

Since A_1 and B_1 are defined from the parameters ξ_1 and Ste, condition (27) will be used to find the value of ξ_1. In this way, it turns out that ξ_1 must be a positive solution of the fourth degree polynomial equation:

$$\begin{aligned} Q_1(z) \quad &:= (1+\text{Ste})^4\left(2\text{Ste}^2+11\text{Ste}+16\right)z^4 \\ &-2(1+\text{Ste})^2\left(6\text{Ste}^2+19\text{Ste}+3\right)z^2 \\ &+3\text{Ste}(1+6\text{Ste}) = 0, \qquad \xi^{\min} < z < \xi^{\max}. \end{aligned} \tag{35}$$

It is easy to see that Q_1 has only two positive roots. In addition:

$$Q_1(\xi^{\min}) = \tfrac{2\text{Ste}^2(2\text{Ste}+3)^2}{(2+\text{Ste})^2} > 0, \tag{36}$$

$$Q_1(\xi^{\max}) = -\tfrac{3\text{Ste}(2\text{Ste}+3)^2}{(3+\text{Ste})^2} < 0, \tag{37}$$

$$Q_1(+\infty) = +\infty. \tag{38}$$

Therefore Q_1 has a unique root in $\left(\xi^{\min}, \xi^{\max}\right)$ and it is given explicitly by

$$\xi_1 = \left(\tfrac{(\text{Ste}+1)^2\left(6\,\text{Ste}^2+19\,\text{Ste}+3\right)-\sqrt{6\,\text{Ste}+1}\left(2\,\text{Ste}^2+5\,\text{Ste}+3\right)}{(\text{Ste}+1)^4\left(2\,\text{Ste}^2+11\,\text{Ste}+16\right)}\right)^{1/2}. \tag{39}$$

All the above analysis can be summarized in the following result:

Theorem 3.1. *The solution to the problem (P1), for a quadratic profile in space, is given by (30) and (31) where the positive constants A_1 and B_1 are defined by (32) and (33) respectively and ξ_1 is given explicitly by (39).*

As this approximate method is designed as a technique for tracking the location of the free boundary, the comparisons between the approximate solutions and the exact one are made on the free boundary thought the coefficients that characterizes them (Fig.1). Generally for most phase-change materials candidates over a realistic temperature, the Stefan number will not exceed 1 [31]. From this, in order to analyse the accuracy of the approximate solution we compare the dimensionless coefficients ξ_1 with the exact coefficient ξ given by (11) for Ste < 1. Moreover, in Fig.2, we show the temperature profile of the approximate solution and the exact one at $t = 10s$, for the parameters Ste $= 0.4, a = 1\ \sqrt{s}/m$ and $\theta_0 = 3°C$.

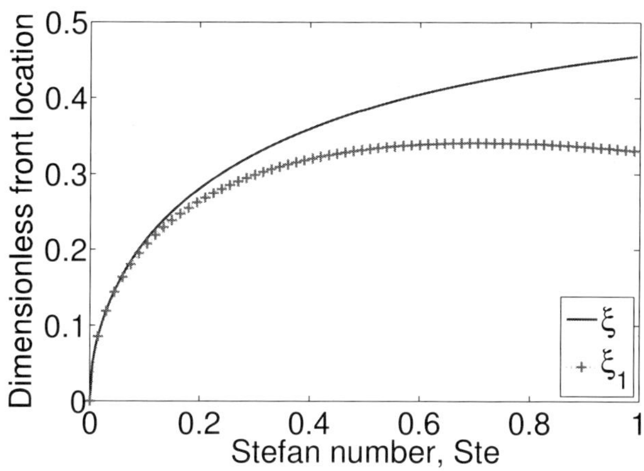

Figure 1. Plot of ξ and ξ_1 against Ste.

3.2. Approximate Solution Using an Alternative of the Heat Balance Integral Method

An alternative method of the classical heat balance integral method proposes to approximate problem (P) through the resolution of the approximate problem defined in Sect. 3, that is:

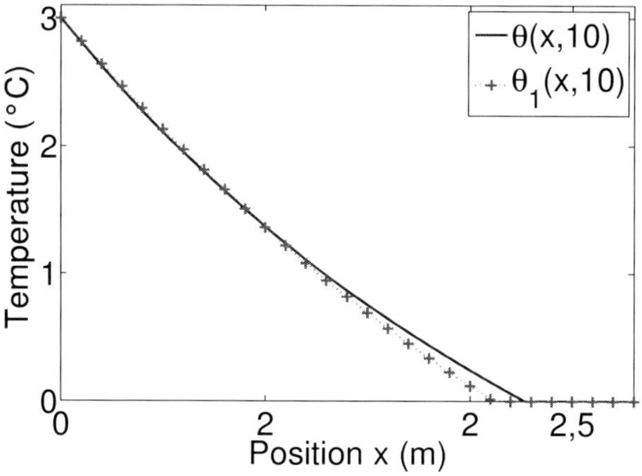

Figure 2. Plot of θ and θ_1 against x at $t = 10$ s for Ste $= 0.4$, $a = 1\ \sqrt{s}/m$, $\theta_0 = 3°C$.

Problem (P2). Find the temperature $\theta_2 = \theta_2(x,t)$ at the liquid region $0 < x < s_2(t)$ and the location of the free boundary $x = s_2(t)$ such that:

$$\frac{d}{dt}\int_0^{s_2(t)} \theta_2(x,t)dx = \frac{-1}{\rho c}\left[\rho\lambda\dot{s}_2(t) + k\left(\theta_0\right)\tfrac{\partial\theta_2}{\partial x}(0,t)\right], \quad 0 < x < s_2(t), \quad (40)$$

$$\theta_2 = \theta_0 > 0, \qquad \text{on } x = 0, \quad (41)$$

$$k\left(\theta_2\right)\frac{\partial\theta_2}{\partial x} = -\rho\lambda\dot{s}_2(t), \qquad \text{on } x = s_2(t), \quad (42)$$

$$\theta_2 = 0, \qquad \text{on } x = s_2(t), \quad (43)$$

$$s_2(0) = 0. \quad (44)$$

The solution of the problem (P2), for a quadratic temperature profile in space, is obtained by

$$\theta_2(x,t) = \theta_0 A_2\left(1 - \frac{x}{s_2(t)}\right) + \theta_0 B_2\left(1 - \frac{x}{s_2(t)}\right)^2, \quad 0 < x < s_2(t),\ t > 0, \quad (45)$$

$$s_2(t) = \frac{2}{a}\xi_2\sqrt{t}, \quad t > 0, \quad (46)$$

where the constants A_2, B_2 y ξ_2 will be determined from the conditions (40), (41) and (42) of the problem (P2). The conditions (43) and (44) are immediately satisfied. From conditions (41) and (42), we obtain:

$$A_2 = \frac{2\xi_2^2}{\text{Ste}}, \tag{47}$$

$$B_2 = 1 - \frac{2\xi_2^2}{\text{Ste}}. \tag{48}$$

As we know, the constants A_2 and B_2 must be positive then we have $0 < \xi_2 < \sqrt{\frac{\text{Ste}}{2}}$.

Moreover, as in the previous problem (P1), the constants A_2 and B_2 are expressed as a function of the parameters ξ_2 and Ste, and using condition (40) the coefficient ξ_2 must be a positive solution of the fourth degree polynomial equation given by:

$$(1+\text{Ste})^2 z^4 + \left(6 + 7\text{Ste} + 5\text{Ste}^2 + \text{Ste}^3\right) z^2 - 3\text{Ste} = 0, \tag{49}$$

for $0 < z < \sqrt{\frac{\text{Ste}}{2}}$.

Then, it is easy to see that the above equation has a unique solution given explicitly by

$$\xi_2 = \left(\frac{-\left(6+7\text{Ste}+5\text{Ste}^2+\text{Ste}^3\right)+\sqrt{\left(6+7\text{Ste}+5\text{Ste}^2+\text{Ste}^3\right)^2+12\text{Ste}(1+\text{Ste})^2}}{2(1+\text{Ste})^2} \right)^{1/2}. \tag{50}$$

All the above analysis can be summarized in the following result:

Theorem 3.2. *The solution to the problem (P2), for a quadratic profile in space, is given by (45) and (46) where the positive constants A_2 and B_2 are defined by (47) and (48) respectively and ξ_2 is given explicitly by (50).*

Fig. 3 shows, for Stefan values up to 1, how the dimensionless coefficient ξ_2, which characterizes the location of the free boundary s_2, approaches the coefficient ξ, corresponding to the exact free boundary s. Moreover, in Fig.4, we show the temperature profile of the approximate solution and the exact one at $t = 10s$ for the parameters Ste $= 0.4$, $a = 1\ \sqrt{s}/m$ and $\theta_0 = 3°C$.

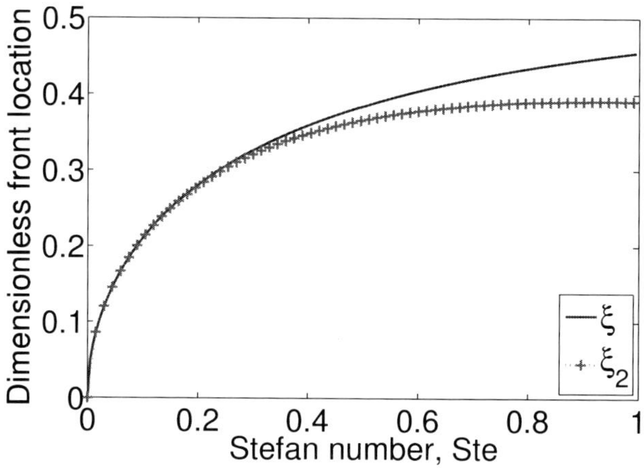

Figure 3. Plot of ξ and ξ_2 against Ste.

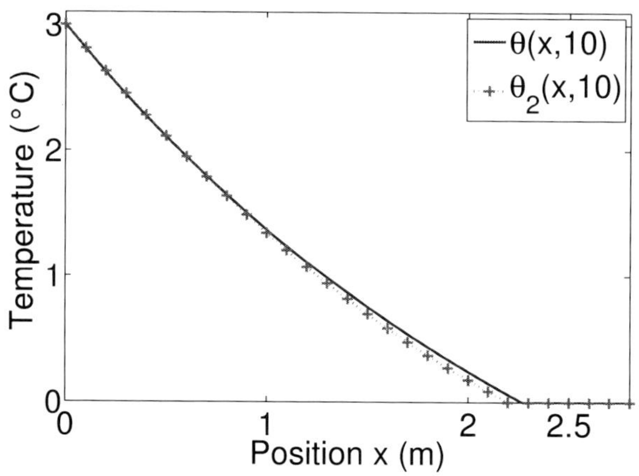

Figure 4. Plot of θ and θ_2 against x at $t = 10$ s for Ste $= 0.4$, $a = 1\ \sqrt{s}/m$, $\theta_0 = 3°C$.

3.3. Approximate Solution Using a Refined Balance Heat Integral Method

The refined integral method proposes to approximate problem (P) through the resolution of an approximate problem formulated in Section 3, that is:

Problem (P3). Find the temperature $\theta_3 = \theta_3(x,t)$ at the liquid region $0 < x < s_3(t)$ and the location of the free boundary $x = s_3(t)$ such that:

$$\int_0^{s_3(t)} \int_0^x \frac{\partial \theta_3}{\partial t}(\eta,t)d\eta dx = \frac{-\theta_0(1+\text{Ste}) - \frac{\partial \theta_3}{\partial x}(0,t)s_3(t)}{a^2(1+\text{Ste})^2}, \qquad 0 < x < s_3(t), \tag{51}$$

$$\theta_3 = \theta_0 > 0, \qquad \text{on } x = 0 \tag{52}$$

$$k(\theta_3)\frac{\partial \theta}{\partial x} = -\rho\lambda\dot{s}_3(t), \qquad \text{on } x = s_3(t), \tag{53}$$

$$\theta_3 = 0, \qquad \text{on } x = s_3(t), \tag{54}$$

$$s_3(0) = 0. \tag{55}$$

The solution of the problem (P3) for a quadratic temperature profile in space is given by:

$$\theta_3(x,t) = \theta_0 A_3\left(1 - \frac{x}{s_3(t)}\right) + \theta_0 B_3\left(1 - \frac{x}{s_3(t)}\right)^2, \qquad 0 < x < s_3(t),\ t > 0 \tag{56}$$

and the free boundary is obtained of the form:

$$s_3(t) = \frac{2}{a}\xi_3\sqrt{t}, \quad t > 0, \tag{57}$$

where the constants A_3 , B_3 y ξ_3 will be determined from the conditions (51), (52) and (53) of the problem (P3).

From conditions (52) and (53) we obtain:

$$A_3 = \frac{2\xi_3^2}{\text{Ste}}, \tag{58}$$

$$B_3 = 1 - \frac{2}{\text{Ste}}\xi_3^2. \tag{59}$$

As is already know $A_3 > 0$ and $B_3 > 0$, thus we obtain that $0 < \xi_3 < \sqrt{\frac{\text{Ste}}{2}}$. Moreover, since A_3 and B_3 are defined from the parameter ξ_3, condition (51) will be used to find the value of ξ_3. In this way it turns out that ξ_3 must be a positive solution of the second degree polynomial equation:

$$\left(\text{Ste}^3 + 2\text{Ste}^2 + \text{Ste} + 6\right)z^2 + 3\text{Ste}(\text{Ste} - 1) = 0, \qquad 0 < z < \sqrt{\frac{\text{Ste}}{2}}. \tag{60}$$

Then, it is easy to see that the above equation has a unique solution if and only if Ste < 1 which is given explicitly by:

$$\xi_3 = \left(\frac{3\text{Ste}(1-\text{Ste})}{\text{Ste}^3 + 2\text{Ste}^2 + \text{Ste} + 6} \right)^{1/2}. \tag{61}$$

All the above analysis can be summarized in the following result:

Theorem 3.3. *The solution to the problem (P3), for a quadratic profile in space, is given by (56) and (57) where the positive constants A_3 and B_3 are defined by (58) and (59) respectively and ξ_3 is given explicitly by (61).*

Therefore for every Ste < 1, we plot the numerical value of the dimensionless coefficient ξ_3 against the exact coefficient ξ (Fig.5). It can be seen that the refined integral method results in good agreement with the exact solution of the problem (P), only for lower values of Stefan number. Moreover, in Fig.6, we show the temperature profile of the approximate solution and the exact one at $t = 10s$ for the parameters Ste $= 0.4$, $a = 1\ \sqrt{s}/m$ and $\theta_0 = 3°C$.

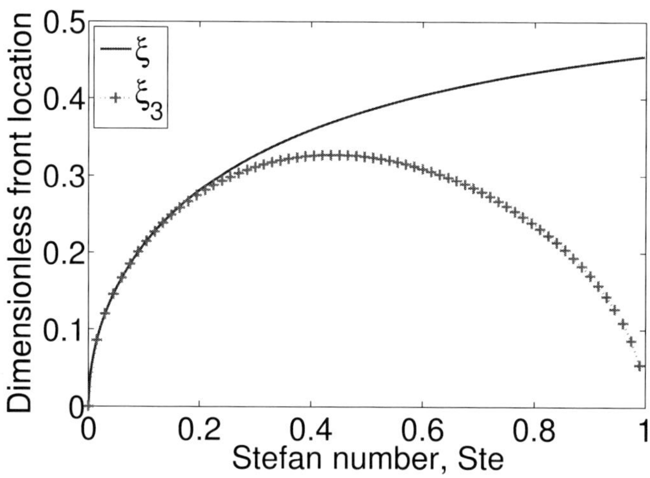

Figure 5. Plot of ξ and ξ_3 against Ste.

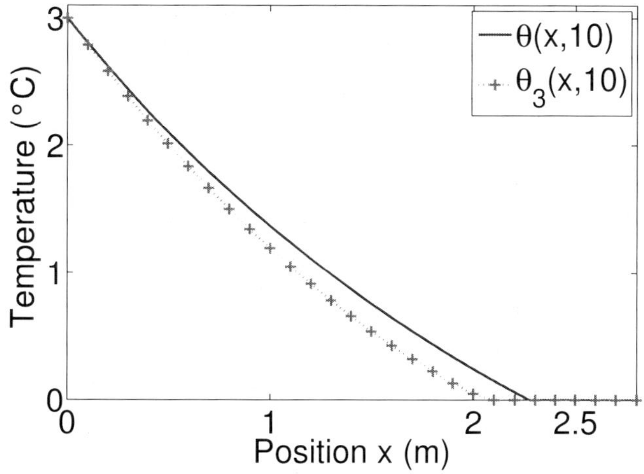

Figure 6. Plot of θ and θ_3 against x at $t = 10$ s for Ste $= 0.4$, $a = 1\ \sqrt{s}/m$, $\theta_0 = 3°C$.

4. Comparisons between Solutions

In the previous sections we have applied 3 different approximate methods (heat balance integral method (HBIM), an alternative of the HBIM and the refined integral method (RIM)) for solving a Stefan problem with a non-linear temperature-dependent thermal conductivity.

For each of this methods, i.e., for each Problem (Pi), $i = 1, 2, 3$ it has been plotted the dimensionless coefficient that characterizes the approximate free front ξ_i versus the coefficient ξ corresponding to the exact moving boundary of problem (P) (Figures 1, 3, 5).

The aim of this section is to present, for different Stefan numbers, the numerical value of the exact coefficient ξ given by (11) and the approximate coefficients ξ_1, ξ_2 and ξ_3 given by the analytical expressions (39), (50) and (61) respectively. Those calculations will allow us not only to compare the approximate solutions with the exact one but also to compare the different approaches between them in order to show which technique gives the best agreement. With that purpose we display in Table 1, for different values of Ste, the exact dimensionless free front ξ, the approximate dimensionless free front ξ_i and the porcentual relative error $E_{\text{rel}}(\xi_i) = 100 \left|\frac{\xi-\xi_i}{\xi}\right|$, $i = 1, 2, 3$.

It may be noticed in Table 1 that the relative error committed in each approximate technique increases when the Stefan number becomes greater reaching the percentages 21%, 14% and 100% for the problems (P1), (P2) and (P3) respectively. From this fact, we study the behaviour of the different approaches for Ste $<<1$ (Table 2). In this case the relative errors for problem (P1), (P2) and (P3) does not exceed 0.5%.

Table 1. Dimensionless free front coefficients and its relative errors

Ste	ξ	textbfξ_1	$E_{\textbf{rel}}(\xi_1)$	ξ_2	$E_{\textbf{rel}}(\xi_2)$	ξ_3	$E_{\textbf{rel}}(\xi_3)$
0.1	0.2099	0.2099	0.037 %	0.2100	0.018 %	0.2100	0.042 %
0.2	0.2805	0.2754	1.803 %	0.2788	0.608 %	0.2763	1.498 %
0.3	0.3262	0.3126	4.194 %	0.3207	1.697 %	0.3112	4.622 %
0.4	0.3593	0.3348	6.809 %	0.3481	3.110 %	0.3258	9.330 %
0.5	0.3846	0.3482	9.470 %	0.3663	4.741 %	0.3244	15.63 %
0.6	0.4046	0.3557	12.09 %	0.3782	6.515 %	0.3091	23.60 %
0.7	0.4209	0.3593	14.63 %	0.3856	8.375 %	0.2802	33.41 %
0.8	0.4343	0.3602	17.07 %	0.3897	10.28 %	0.2364	45.58 %
0.9	0.4457	0.3592	19.41 %	0.3913	12.20 %	0.1709	61.66 %
1.0	0.4554	0.3568	21.63 %	0.3911	14.11 %	0	100.0 %

Table 2. Dimensionless free front coefficients and its relative errors

Ste	ξ	ξ_1	$E_{\textbf{rel}}(\xi_1)$	ξ_2	$E_{\textbf{rel}}(\xi_2)$	ξ_3	$E_{\textbf{rel}}(\xi_3)$
0.01	0.0702	0.0703	0.142 %	0.0703	0.037 %	0.0703	0.075 %
0.02	0.0987	0.0989	0.241 %	0.0988	0.066 %	0.0988	0.135 %
0.03	0.1201	0.1205	0.302 %	0.1202	0.086 %	0.1203	0.178 %
0.04	0.1378	0.1382	0.329 %	0.1379	0.099 %	0.1381	0.206 %
0.05	0.1531	0.1536	0.326 %	0.1532	0.103 %	0.1534	0.219 %
0.06	0.1666	0.1671	0.296 %	0.1668	0.101 %	0.1670	0.215 %
0.07	0.1789	0.1793	0.242 %	0.1790	0.090 %	0.1792	0.196 %
0.08	0.1901	0.1904	0.167 %	0.1902	0.073 %	0.1904	0.160 %
0.09	0.2004	0.2005	0.073 %	0.2005	0.049 %	0.2006	0.109 %

In Figure 7 we present a comparison of the absolute errors of the approximate temperatures given by $E_{\text{abs}}(\theta_i(x,t)) = |\theta(x,t) - \theta_i(x,t)|$, $i = 1,2,3$

against the position x, at $t = 10s$, Ste $= 0.4$, $a = 1\sqrt{s}/m$ and $\theta_0 = 3°C$.

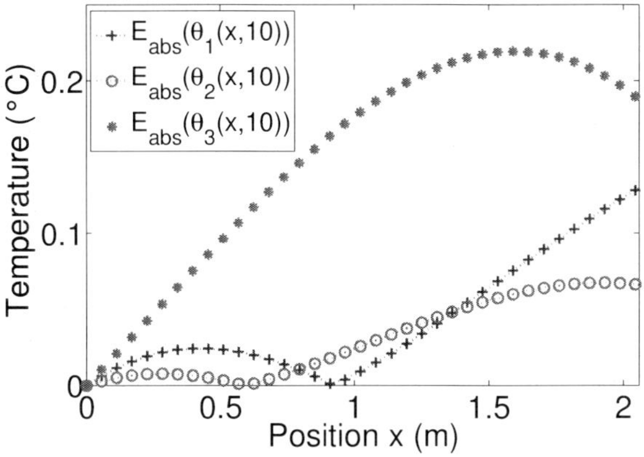

Figure 7. Temperatures absolute errors against x at $t = 10s$ for Ste $= 0.4$, $a = 1\sqrt{s}/m$ and $\theta_0 = 3°C$.

Remark 4.1. *In order to compare the absolute errors of the different approaches in a common domain, in Figure 7, we plot up to* $x = s_3(10) = \min\{s_1(10), s_2(10), s_3(10)\}$.

Acknowledgments

The present work has been partially sponsored by Projects PIP No. 112-20150100275CO CONICET-Univ. Austral and PICTO Austral 2016 No. 0090, Rosario (Argentina).

Conclusion

In this chapter it was provided an overview of the popular approaches such as HBIM, RIM, for the case of a one-dimensional one-phase Stefan problem (P) with a non-linear temperature-dependent thermal conductivity as the novel feature.

It must be emphasized that the fact of having the exact solution of problem (P) has allowed us to measure the accuracy of the approximate techniques applied throughout this chapter. Comparisons with known solution have been made in all cases and all solutions have been presented in graphical form.

It has been observed that as the Stefan number increases, the coefficients that characterizes the free approximate boundaries move away from the exact one. However, for Ste $<< 1$, the three approaches commit a relative error that does not exceed 0.5%.

In all the analysed cases, it could be concluded that the alternative technique of HBIM given by problem (P2) is significantly more accurate than the others.

References

[1] Alexiades V. and SolomonA. D., *Mathematical Modelling of Melting and Freezing Processes*, Hemisphere-Taylor: Francis, Washington, 1993.

[2] Cannon J. R., *The one-dimensional heat equation*, Addison-Wesley: Menlo Park, California, 1984.

[3] Gupta S. C., *The classical Stefan problem. Basic concepts, modelling and analysis*, Elsevier: Amsterdam, 2003.

[4] Lunardini V. J., *Heat transfer with freezing and thawing*, Elsevier: London, 1991.

[5] Tarzia D. A., A bibliography on moving-free boundary problems for the heat-diffusion equation, The Stefan and related problems, *MAT-Serie A* **2**, 1-297 (2000).

[6] Tarzia D. A., Explicit and approximated solutions for heat and mass transfer problem with a moving interface, Chapter 20, In: *Advanced Topics in Mass Trasnfer*, M. El-Amin (Ed). InTech Open Access Publisher, Rijeka, Croatia, 2011, pp. 439–484.

[7] Goodman T. R., The heat balance integral methods and its application to problems involving a change of phase, *Transactions of the ASME* **80**, 335-342 (1958).

[8] Wood A. S., A new look at the heat balance integral method, *Applied Mathematical Modelling* **25**, 815-824 (2001).

[9] Hristov J., The heat-balance integral method by a parabolic profile with unspecified exponent: analysis and benchmark exercises, *Thermal Science* **13**, 27-48 (2009).

[10] Hristov J., Research note on a parabolic heat-balance integral method with unspecified exponent: An Entropy Generation Approach in Optimal Profile Determination, *Thermal Science* **13**, 49-59 (2009).

[11] Mitchell S., Applying the combined integral method to one-dimensional ablation, *Applied Mathematical Modelling* **36**, 127-138 (2012) .

[12] Mitchell S. and Myers T., Improving the accuracy of heat balance integral methods applied to thermal problems with time dependent boundary conditions, *Int. J. Heat Mass Transfer* **53**, 3540-3551 (2010).

[13] Mitchell S. and Myers T., Application of Heat Balance Integral Methods to One-Dimensional Phase Change Problems, *Int. J. Diff. Eq.* **2012**, 1-22 (2012).

[14] F. Mosally, Wood A. and Al-Fhaid A., An exponential heat balance integral method, *Applied Mathematics and Computation* **130**, 87-100 (2002).

[15] Barry D. A. and Sander G. C., Exact solutions for water infiltration with an arbitrary surface flux or nonlinear solute adsorption, *Water Resources Research.* **27**, 2667-2680 (1991).

[16] Bluman G. and Kumei S., On the remarkable nonlinear diffusion equation, *J. Math Phys.* **21**, 1019-1023 (1980).

[17] Broadbridge P., Non-integrability of non-linear diffusion-convection equations in two spatial dimensions, *J. Phys. A: Math. Gen.* **19**, 1245-1257 (1986).

[18] Broadbridge P., Integrable forms of the one-dimensional flow equation for unsaturated heterogeneous porous media, *J. Math. Phys.* **29**, 622-627 (1988).

[19] Fokas A. S. and Yortsos Y. C., On the exactly solvable equation $S_t = \left[(\beta S+\gamma)^{-2} S_x\right]_x + \alpha(\beta S+\gamma)^{-2} S_x$ occurring in two-phase flow in porous media, *SIAM J. Appl. Math.* **42**, 318-331 (1982).

[20] Knight J. H. and Philip J. R., Exact solutions in nonlinear diffusion, *J. Engrg. Math.* **8**, 219-227 (1974).

[21] Munier A., Burgan J. R., Gutierrez J., Fijalkow E. and Feix M. R., Group transformations and the nonlinear heat diffusion equation, *SIAM J. Appl. Math.* **40**, 191-207 (1981).

[22] Philip R., General method of exact solution of the concentration-dependent diffusion equation, *Australian J. Physics* **13**, 1-12 (1960).

[23] Rogers C., Application of a reciprocal transformation to a two-phase Stefan problem, *J. Phys. A: Math. Gen.* **18**, 105-109 (1985).

[24] Sander G. C., Cunning I. F., Hogarth W. L. and Parlange J. Y., Exact solution for nonlinear nonhysteretic redistribution in vertical solid of finite depth, *Water Resources Research* **27**, 1529-1536 (1991).

[25] Tritscher P. and Broadbridge P., A similarity solution of a multiphase Stefan problem incorporating general non-linear heat conduction, *Int. J. Heat Mass Transfer* **37**, 2113-2121 (1994).

[26] Natale M. F. and Tarzia D. A., Explicit Solutions for a One-phase Stefan Problem with Temperature-dependent Thermal Conductivity, *Bolletino U.M.I.* **8**, 79-99 (2006).

[27] Sadoun N., Si-Ahmed E. K. and Colinet P., On the refined integral method for the one-phase Stefan problem with time-dependent boundary conditions, *Applied Mathematical Modelling* **30**, 531-544 (2006).

[28] Fabre A. and Hristov J., On the integral-balance approach to the transient heat conduction with linearly temperature-dependent thermal diffusivity, *Heat Mass Transfer* **53**, 177-204 (2017).

[29] Hristov J., An approximate analytical (integral-balance) solution to a nonlinear heat diffusion equation, *Thermal Science* **19**, 723-733 (2015).

[30] Hristov J., Integral solutions to transient nonlinear heat (mass) diffusion with a power-law diffusivity: a semi-infinite medium with fixed boundary conditions, *Heat Mass Transfer* **52**, 635-655 (2016).

[31] Solomon A. D., An easily computable solution to a two-phase Stefan problem, *Solar energy* **33**, 525-528 (1979).

In: Heat Conduction
Editors: J. Hristov and R. Bennacer
ISBN: 978-1-53614-673-8

Chapter 2

APPLICATION OF VARIATIONAL INTEGRAL METHOD TO ANALYZE VARIETY OF REWETTING PROBLEMS

Manish Kumar Agrawal[1,*] ***and Santosh Kumar Sahu***[2,†]
[1]School of Engineering Sciences, Mahindra Ecole Centrale, Hyderabad, India
[2]Discipline of Mechanical Engineering, School of Engineering, Indian Institute of Technology Indore, Simrol, Indore, India

Abstract

Variational integral method has been employed to analyze variety of rewetting problems. This includes the basic two region rewetting model that assumes a constant heat transfer coefficient in the wet region and adiabatic condition in the dry region ahead of wet front. Subsequently the analysis has been extended to include the effect of precursory cooling, variation in heat transfer in multiple step functions, exponential functions and variation in property. Based on the analysis, closed form solution is obtained for the temperature field and rewetting velocity.

Keywords: variational integral method, conduction, rewetting, Ritz profile, Peclet number

*Corresponding Author Email: agrwal.manish85@gmail.com.
†E-mail address: santosh.sahu04@gmail.com.

1. Introduction

Quenching is a technique used to cool sufficiently hot objects in various industrial applications. When liquid is brought in contact with sufficiently hot surface, it cannot wet the hot surface immediately. A vapour blanket is formed on the solid surface that prevents the contact between hot surface and liquid. Consequently, the heat transfer from the hot surface reduces due to poor conduction of heat through the vapour layer. If the process is allowed to continue, the hot surface reaches a temperature at which the vapour film collapses and the liquid wet the solid surface. This process is termed as rewetting. This means re-establishment of hot surface-liquid contact. Rewetting is considered as an important process during quenching in view of heat transfer rate.

In general two different approaches have been followed to analyze the phenomenon of rewetting. One can consider the rewetting process as conduction controlled or hydrodynamically controlled for the analysis. In the case of rewetting by top flooding, axial conduction plays a dominant role. In such a case, conduction-controlled rewetting models could represent the physical phenomena reasonably well while reducing the complexity of the problem. Various analytical and semi-analytical techniques have been used to solve the conduction equation. These include analytical techniques such as separation of variables method, Winer-Hopf technique, and heat balance integral method (HBIM). Some of the techniques compute the solution by truncating the series of complex function by employing Fourier expansion. The truncation of infinte series of a function may lead to errors in the function. On the contrary, various numerical techniques have been used to solve the conduction equation such as finite difference technique, finite element method, and implicit isotherm migration technique. The accuracy of the solution obtained by numerial technique mainly depends on the choice of structural meshing. Therefore, in order to obtain the solution with acceptable accuracy, a dense discritization mesh is usually used, which is computationally inefficient and time consuming as well. The main objective of the analytical models is to predict rewetting velocity, temperature distribution and rewetting temperature. This includes mainly the analytical and numerical solution of various rewetting problems comprising different geometry, different convective boundary condition, heat generation, property variation and coupling effect of coolant flow. The variational integral method proposed by Biot [1] has mostly been used to solve problems of elasticity and applied sciences. Earlier Sparrow and Siegel [2] used the variational integral method to evaluate the fully developed velocity profile and temperature distribution in different duct geometries. Later on, Arpaci [3] applied the same method to solve variety of conduction problems. Although the techniqe is simple, it provides results with reasonable accuracy. In the present chapter variatinoal integral method (VIM) is employed to analyze the phenomenon of rewetting in various cases namely, two region rewetting model, rewetting model with precursory cooling and rewetting model with temperature dependent thermal conductivity.

1.1. Method of Variational Calculus

Variational calculus includes finding a function $y(x)$ such that a definite integral whose integrand, is a function of $y(x)$ assume an extreme value. For example the differential length of a curve (Fig. 1), $y = y(x)$ connecting two different points P_1 and P_2 can be expressed as:

$$ds = (dx^2 + dy^2)^{\frac{1}{2}} \tag{1}$$

which can be rearranged as:

$$ds = (1 + y'^2)^{\frac{1}{2}} dx \tag{2}$$

where y'=$\frac{dy}{dx}$
The total length of this curve can be written as:

$$S = \int_a^b (1 + y'^2)^{\frac{1}{2}} dx \tag{3}$$

At this juncture, one needs to find the minimum value of this function.

To find extreme values of an Integral

$$I = \int_a^b F(x, y, y')dx \tag{4}$$

where $F(x, y, y')$ is a function known from the statement of the problem, While y is the unknown to be determined such that the integral becomes stationary. A function defined by a definite integral whose integrand itself depends on functions is said to be a *functional*. The integrand of this integral may also be called

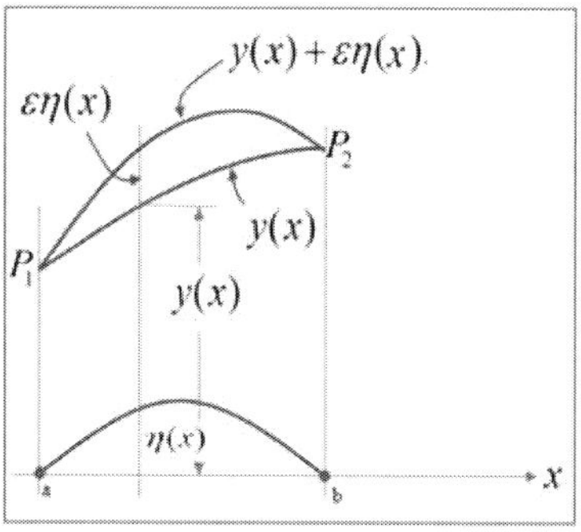

Figure 1. Variation of the dynamic permeability with porosity.

a functional when the independent variable x is considered fixed but the dependent variable y is varied. The integal of (4) takes on different values along different paths connecting the fixed points P_1,P_2. We have to find the particular curve $y(x)$ which makes the integral *minimum*.

If it is assumed that $y(x)$ actually minimizes this integral, any function in the neighborhood of $y(x)$ can be represented by the form $y(x) + \varepsilon\eta(x)$, where $\eta(x)$ is a continuously differentiable function that vanishes at the ends of the interval, and ε is a parameter as shown in Figure 1.1. Then in terms of these functions, the integral of (4) can be written as:

$$I(\varepsilon) = \int_a^b F(x, y + \varepsilon\eta, y' + \varepsilon\eta')dx \tag{5}$$

Which, according to the assumption about $y(x)$, assumes its minimum value when $\varepsilon = 0$.

Since in the (I,ε)-plane, $I(\varepsilon)$ has a minimum for $\varepsilon = 0$, according to differential calculus (I,ε)satisfies the necessary condition for extremum at this point, namely

$$\frac{dI(\varepsilon)}{d\varepsilon} = 0; \quad when, \varepsilon = 0 \tag{6}$$

Now, defining

$$F(\varepsilon) = F(x, y + \varepsilon\eta, y' + \varepsilon\eta') \tag{7}$$

And assuming that the order of integration and differentiation can be interchanged, we combine (5), (6) and (7) to give

$$\frac{dI(\varepsilon)}{d\varepsilon} = \int_a^b \frac{dF(\varepsilon)}{d\varepsilon} dx \tag{8}$$

Since for a fixed value of x the derivative of (7) with respect to ε is

$$\frac{dF(\varepsilon)}{d\varepsilon} = \frac{\partial F(\varepsilon)}{\partial y + \varepsilon\eta}\eta + \frac{\partial F(\varepsilon)}{\partial y' + \varepsilon\eta'}\eta' \tag{9}$$

Noting that

$F(\varepsilon)\Big|_{\varepsilon=0} = F$ and utilizing (5), (7), (8) and (9); Eq.(6) can be written as:

$$\left.\frac{dI(\varepsilon)}{d\varepsilon}\right|_{\varepsilon=0} = \int_a^b \left(\frac{\partial F}{\partial y}\eta + \frac{\partial F}{\partial y'}\eta'\right) dx = 0$$

Integrating the second term of this integral by parts, we obtain

$$\int_a^b \left[\frac{\partial F}{\partial y} - \frac{d}{dx}\left(\frac{\partial F}{\partial y'}\right)\right] \eta dx + \left.\frac{\partial F}{\partial y'}\eta\right|_a^b = 0 \tag{10}$$

Since, by definition, η vanishes at the ends of the interval, the second term of (10) is zero. Hence

$$\int_a^b \left[\frac{\partial F}{\partial y} - \frac{d}{dx}\left(\frac{\partial F}{\partial y'}\right)\right] \eta dx = 0 \tag{11}$$

The integral of (11) vanishes for any $\eta(x)$ when

$$\frac{\partial F}{\partial y} - \frac{d}{dx}\left(\frac{\partial F}{\partial y'}\right) = 0 \tag{12}$$

Equation (12) is the Euler equation associated with the variational problem given by (4). Thus the condition necessary for $y(x)$ to minimize (or maximize) the integral of (4) is that $F(x, y, y')$ must satisfy the corresponding Euler equation. In general, variational method is used by utilizing either the Sparrow and Siegel approach [2] or Arpaci approach [3] to solve variety of problems. Different approaches as reported by the researchers are elaborated below.

1.1.1. Sparrow and Siegel [2] Approach

The Eulers equation, valid for the governing equation of the physical problem can be written in the following form [2]:

$$\frac{\partial}{\partial x}\left(\frac{\partial F}{\partial \theta'}\right) - \frac{\partial F}{\partial \theta} = 0 \tag{13}$$

where, the functional $F = f(x, \theta, \theta', \theta'')$ is to be selected in such a way that, when it is inserted into the Euler's equation (13) leads to the governing differential equation of the physical problem. Once the functional $F = f(x, \theta, \theta', \theta')$ is selected, based on the boundary condition of the physical problem, suitable guess temperature profile is approximated and integral of the variational formulation is obtained. Further, minimization of the integral provides various constants associated with guess temperature profile. A detailed description of the Sparrow approach is explained in section 1.2.1.

1.1.2. Arpaci [3] Approach

Unlike the Sparrow and Siegel [2], Arpaci [3] proposes the *mathematical statement of physical problem* (governing differential equation) as the *functional* $F = f(x, \theta, \theta', \theta'')$ *for the variational formulation*. In a view of this the differential equation of the physical problem is considered to be the Euler equation associated with its desirable variational formulation. It is to be noted that *unlike the differential formulation, that involves the governing differential equation and the boundary condition separately, the variational formulation includes the boundary conditions in the governing variational equation. Further, it is likely that a problem may yield different variational formulation, depending upon the boundary conditions. Success of variational formulation depends upon approximate geometrical profile which satisfies the prescribed boundary condition of a given problem.* However, among these profiles there is only one physically possible profile which, by satisfying the appropriate Euler equation, makes the functional of the corresponding variational problem stationary. It is often difficult to guess this exact profile. However, some approximate methods, such as *Ritz* and *Kantorvich* methods available in the literature, provides a basis for the approximation of guess profile for which the functional $F = f(x, \theta, \theta', \theta'')$ has minimum value [3]. The methodology to choose the approximate guess profile is explained in the subsequent sections. Here, we have shown the application of variational method to solve several rewetting problems.

1.2. Application of VIM to Solve Rewetting Problems

It may be noted that the method of variational principle has been extensively used to solve problem of elasticity and applied sciences [2]. Earlier, Sparrow and Siegel [2] used the variational method to evaluate the fully developed velocity profile and temperature distribution in different geometries. Their results exhibited an excellent agreement with the exact solution. Later on, Arpaci [3] applied the same method to solve variety of conduction problems. In the present study an effort has been made to employ this technique to analyze variety of rewetting problems involving various convective boudary conditions, property varaition, precrusory cooling. Efforts have been made to utilize both the approaches [2, 3] to solve the conduction equation to analyze the rewetting behavior of hot surface.

1.2.1. Two Region Rewetting Model

It is observed that during top flooding rewetting, axial conduction plays a dominant role [4, 10]. Therefore, a conduction model is adopted for analyzing the rewetting phenomenon. Figure 2a schematically depicts the top flow rewetting of a one-dimensional slab of infinite length.

Following assumptions are made for the analysis:

1. The wet front velocity is assumed to be constant during its propagation. The end effects are neglected. Therefore, the problem can be reduced to a quasi-steady one.

2. Based on the above assumption, one can conceive of a two region (wet and dry) conduction problem with a sharp temperature gradient at the interface.

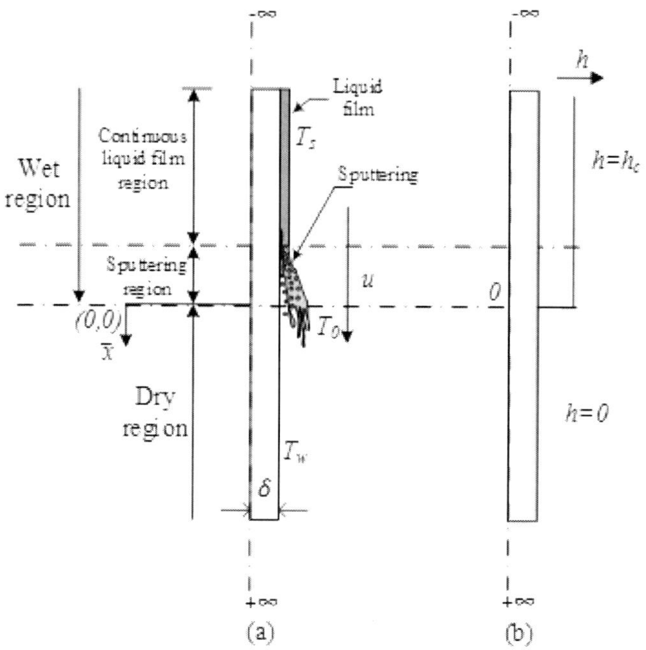

Figure 2. (a) Schematic of one-dimensional hot object (b) variation of heat transfer coefficient along the axial direction [6].

3. The constant heat transfer coefficient is assumed in the wet region while an adiabatic condition is considered in the dry region ahead of quench front (i.e. precursory cooling is neglected) Figure 2b. This is justified for lower coolant flow rates, however, at higher coolant flow rates; neglecting precursory cooling may underpredict the rewetting velocity.

4. In the liquid region behind the wet front the surface temperature approaches liquid saturation temperature (T_s). The temperature of the surrounding medium for the dry region is assumed to be equal to (T_w).

5. Suitable values of rewetting temperature and heat transfer coefficients are taken as input parameters to solve the conduction equation.

6. Other factors such as: surface finish, surface roughness is not considered in the analysis.

Quasi-steady assumption

The one-dimensional conduction equation for the solid in Cartesian coordinate system (Figure 2) can be written as:

$$K\frac{\partial^2 T}{\partial \bar{x}^2} - \frac{h}{\delta}(T - T_s) = \rho C\frac{\partial T}{\partial t} \qquad (-\infty < x < +\infty) \tag{14}$$

Where K,ρ,C represents thermal conductivity, density and specific heat of the slab material respectively. In order to convert the above transient equation into a quasi-steady one, the stationary coordinate (X) have been transformed into the moving $(\bar{x})$ coordinate system (Figure 3).

The heat transfer phenomena during a moving heat source/sink are of great importance in several engineering applications namely, welding, continuous casting, flame hardening and quenching. In such a case, the constant heat source/sink (wet front) moves in the axial direction with a uniform velocity with respect to a stationary coordinate system (X) (Figure 3). One can assume that an observer stationed at a point on the moving $\bar{x} - axis$ will find no noticeable change in the temperature of his surroundings. This condition termed as $apparent$ steady state temperature, is represented mathematically as $\frac{\partial T}{\partial t} = 0$ in the moving coordinate system [7].

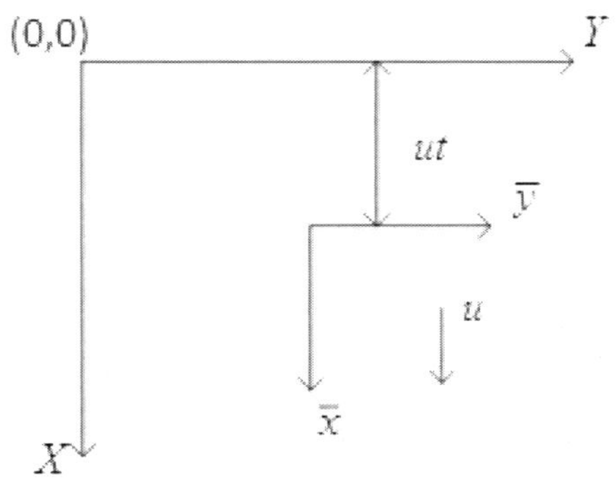

Figure 3. Moving coordinate system $(\bar{x}) = X - ut$.

The transformation of Eq. (14) into moving coordinate system can be done by defining new variables as $\bar{x} = X - ut$ and $\tau = t$. This can further be used to transform (14) as:

$$\frac{\partial T}{\partial X} = \frac{\partial T}{\partial \bar{x}}\frac{\partial \bar{x}}{\partial X} + \frac{\partial T}{\partial \tau}\frac{\partial \tau}{\partial X} = \frac{\partial T}{\partial \bar{x}} \qquad \left(\frac{\partial \bar{x}}{\partial X} = 1 \quad \text{and} \quad \frac{\partial \tau}{\partial X} = 0\right) \tag{15}$$

Similarly,

$$\frac{\partial^2 T}{\partial X^2} = \frac{\partial^2 T}{\partial \bar{x}^2} \tag{16}$$

also the transient term in (14) can be expressed as:

$$\frac{\partial T}{\partial t} = \frac{\partial T}{\partial \bar{x}}\frac{\partial \bar{x}}{\partial t} + \frac{\partial T}{\partial \tau}\frac{\partial \tau}{\partial t} = -u\frac{\partial T}{\partial \bar{x}} + \frac{\partial T}{\partial \tau} \tag{17}$$

as $\frac{\partial \bar{x}}{\partial t} = -u \qquad and \qquad \frac{\partial \tau}{\partial t} = 1$

Substituting partial derivatives of (15), (16), (17) into (14), one can get:

$$K\frac{\partial^2 T}{\partial \bar{x}^2} - \frac{h}{\delta}(T - T_s) = \rho C\left(-u\frac{\partial T}{\partial \bar{x}} + \frac{\partial T}{\partial \tau}\right) \tag{18}$$

Using Rosenthal approximation [7], $\frac{\partial T}{\partial \tau} = 0$ and hence one can obtain:

$$K\frac{\partial^2 T}{\partial \bar{x}^2} - \frac{h}{\delta}(T - T_s) + \rho C u\frac{\partial T}{\partial \bar{x}} = 0 \tag{19}$$

Equation (19) is identified as governing equation for the Cartesian geometry in quasi-steady state condition.

One-dimensional formulation

The one-dimensional governing equation of a hot solid in the quasi-steady condition is expressed in (19). The following normalized variables are defined:

$$x = \frac{\bar{x}}{\delta}, \quad Pe = \frac{u\delta\rho C}{K}, \quad Bi = \frac{h\delta}{K}, \quad \theta = \frac{T - T_s}{T_0 - T_s}, \quad \theta_1 = \frac{T_w - T_0}{T_0 - T_s} \tag{20}$$

Utilizing (20), the energy equation (19) is transformed into following form:

$$\frac{d^2\theta}{dx^2} + Pe\frac{d\theta}{dx} - mBi\theta = 0 \begin{cases} m = 1, & wetregion \\ m = 0, & dryregion \end{cases} \tag{21}$$

Subjected to boundary condition given by:

$$B.C. = \begin{cases} \theta(-\infty) = 0; \theta(0) = 1; & (-\infty < x \leq 0) \\ \theta(0) = 1; \theta(+\infty) = 1 + \theta_1; & (0 \leq x \leq +\infty) \end{cases} \tag{22}$$

Sparrow and Siegel [2] approach

The Euler's equation, valid for the governing equation (21), can be written in the following form [2]:

$$\frac{\partial}{\partial x}\left(\frac{\partial F}{\partial \theta'}\right) - \frac{\partial F}{\partial \theta} = 0 \tag{23}$$

where the functional $F = f(x, \theta, \theta', \theta'')$ is to be selected in such a way that, when it is inserted in to the Euler's equation (23) leads to the governing differential equation (21). We have tried different functional forms in (23) and the ones satisfying the above criteria are expressed as:

$$F = \begin{cases} e^{Pex}\left[\left(\frac{d\theta}{dx}\right)^2 + Bi\theta^2\right], & -\infty < x \leq 0 \\ e^{Pex}\left(\frac{d\theta}{dx}\right)^2, & 0 \leq x < +\infty \end{cases} \tag{24}$$

After defining the functional $F = f(x, \theta, \theta', \theta'')$, the variational integral valid for both wet and dry region is defined as:

$$I = \begin{cases} \int_{-\infty}^{0} F dx, & -\infty < x \leq 0 \\ \int_{0}^{+\infty} F dx, & 0 \leq x < +\infty. \end{cases} \tag{25}$$

At this junction, it is necessary to select a suitable guess temperature profile that satisfies the boundary conditions of the problem. In view of this, we have tried a guess profile for both wet and dry region containing different constants and are expressed as:

$$\theta(x) = \begin{cases} M_1 e^{-a_0 x} + M_2 e^{b_0 x}, & -\infty < x \leq 0 \\ M_3 + M_4 e^{-c_0 x}, & 0 \leq x < +\infty. \end{cases} \tag{26}$$

where $M_i, i = 1to4, a_0, b_0$ *and* c_0 are positive constants. In the variational principle, the variational integral (I) is to be minimized to evaluate constants approximated in the guess temperature profile. This is expressed mathematically as:

$$\frac{\partial I}{\partial E} = 0 \tag{27}$$

where $E = a_0, b_0, c_0$ Using boundary condition given in (22), and (24)-(27) yields:

$$\theta(x) = \begin{cases} e^{\frac{Pe}{2}\left[\left(1+\frac{4Bi}{Pe^2}\right)^{\frac{1}{2}}-1\right]x}, & -\infty < x \leq 0 \\ (1+\theta_1) - \theta_1 e^{-Pex}, & 0 \leq x < +\infty \end{cases} \tag{28}$$

Equation 28 represents the temperature distribution in the hot solid for both wet and dry region. It may be noted that the wet front velocity can be calculated by applying heat flux continuity at the interface as:

$$\left(\frac{d\theta}{dx}\right)_{x=o^+} = \left(\frac{d\theta}{dx}\right)_{x=o^-} \tag{29}$$

Which yields,

$$\frac{\sqrt{Bi}}{Pe} = \sqrt{\theta_1(1+\theta_1)} \tag{30}$$

Based on the analysis a three parametric relation between various modeling parameters such as: Bi,Pe and θ_1 is obtained (30) and is exactly same with the previous results reported by various researchers [5, 8, 9, 10].

Arpaci [3] approach

Unlike Sparrow and Siegel [2], Arpaci [3] proposes the governing differential equation as the functional $F = f(x, \theta, \theta', \theta'')$ for the analysis. Following Arpaci [3] approach, the variational integral valid for the present configuration can be expressed as:

$$I = \begin{cases} \int\limits_{-\infty}^{0} \left(\frac{d^2\theta}{dx^2} + Pe\frac{d\theta}{dx} - Bi\theta\right)dx, & -\infty < x \leq 0 \\ \int\limits_{0}^{+\infty} \left(\frac{d^2\theta}{dx^2} + Pe\frac{d\theta}{dx}\right)dx, & 0 \leq x < +\infty \end{cases} \tag{31}$$

It may be noted that the success of variational integral method depends on the assumed guess profile. However, there is hardly any guideline to select the best profile. Previously polynomial guess functions of different orders were used to analyze the heat transfer and fluid flow problems [2]. It may be noted that, in the present problem, the temperature decays (or increases slowly) away from the origin. In view of this, a hybrid profile consisting of different orders of polynomial and exponential function is considered as below.

$$\theta(x) = \begin{cases} M_5 + M_6 e^x (1 + a_1 x), & -\infty < x \leq 0 \\ N_1(1 - e^{-x}) + e^{-x}(1 + b_1 x), & 0 \leq x < +\infty \\ M_7 + M_8 e^x (1 + a_2 x + a_3 x^2), & -\infty < x \leq 0 \\ N_2(1 - e^{-x}) + e^{-x}(1 + b_2 x + b_3 x^2), & 0 \leq x < +\infty \end{cases} \tag{32}$$

where $M_i(i = 5\ to\ 8), N_j(j = 1\ to\ 2), a_k(k = 1\ to\ 3)$ and $b_k(k = 1\ to\ 3)$ are positive constants. At this juncture, the variational integral (I) is minimized to evaluate the constants and is expressed as:

$$\frac{\partial I}{\partial G} = 0 \tag{33}$$

where, $G = a_1, a_2, a_3, b_1, b_2, b_3$
By using eq. (31-33) and boundary conditions (22) one gets:

$$\theta(x) = \begin{cases} \text{First order approximation} \begin{cases} e^x(1 + a_1 x), & -\infty < x \leq 0 \\ (1 + \theta_1)(1 - e^{-x}) + e^{-x}(1 + b_1 x), & 0 \leq x < +\infty \end{cases} \\ \text{Second order approximation} \begin{cases} e^x(1 + a_2 x + a_3 x^2), & -\infty < x \leq 0 \\ (1 + \theta_1)(1 - e^{-x}) + e^{-x}(1 + b_2 x + b_3 x^2), & 0 \leq x < +\infty \end{cases} \end{cases} \tag{34}$$

where,

$$a_1 = -\frac{1 + Pe - Bi}{1 + Bi}, \qquad b_1 = \theta_1(Pe - 1),$$
$$a_2 = \frac{2Pe^2 + 6Bi^2 - (8PeBi + 2 + 4Bi)}{3Bi^2 + 10Bi - (29 + 29Pe + 7Pe^2)}, \qquad b_2 = \frac{2\theta_1(Pe^2 - 1)}{Pe^2 + 3}$$
$$a_3 = \frac{2Pe^2 + 2Bi^2 + 4Pe + 2 - (4PeBi + 4Bi)}{3Bi^2 + 10Bi - (7Pe^2 + 29Pe + 29)}, \qquad b_3 = -\frac{2\theta_1(1 - Pe)^2}{Pe^2 + 3} \tag{35}$$

By applying Arpaci [3] approach, two different guess profiles are adopted to evaluate the temperature distribution in the hot solid. Based on the analysis the temperature profile for both wet and dry region is expressed as a function of Biot number, Peclet number and axial distance away from wet front. Using Eq. (29) *and* (34), (35) one can evaluate the wet front velocity as:

$$\theta_1 = \begin{cases} \dfrac{2Bi - Pe}{Pe(1+Bi)}, & \text{First order approximation} \\ \left(\dfrac{2+4Bi-6Bi^2-2Pe^2+8PeBi}{29+10Bi+7Pe^2+29Pe-3Bi^2}+1\right)\left(\dfrac{3+Pe^2}{1+3Pe^2}\right), & \text{Second order approximation} \end{cases} \tag{36}$$

Eq. (36) represents a three parametric relationship between various modeling parameters such as: dry wall temperature (θ_1), Peclet number (Pe), and Biot number (Bi). Similar expressions have been obtained by previous researchers as well. However, different functional forms were obtained in the analysis [4, 8].

1.2.2. The Effect of Precursory Cooling on Rewetting of Hot Vertical

It is observed that when water is sprayed at lower flow rates from the top, the water flows down in the form of film and cools the hot object. However, in the case of higher flow rates, droplet spray arises due to instabilities in the upstream water layer. As a result, part of the coolant sputters away from the wet front. The sputtered droplets impact the downstream surface where dryout has occurred and provide cooling to dry wall by droplet vaporization. This phenomenon is termed as precursory cooling. Droplet vaporization results from convective heat transfer and radiative heat transfer across a thin vapor layer separating the droplet from the surface. It is therefore essential to include the precursory cooling in the model to analyze the rewetting of hot surface at higher flow rates. Several researchers have proposed the rewetting models with precursory cooling. These authors have considered a constant heat transfer coefficient [11, 12], variation in heat transfer coefficient [10, 13] or variation in the heat flux [14, 15, 16] in the dry region ahead of wet front to model the effect of precursory cooling. In this model an exponentially decaying heat transfer coefficient of the form $h(\bar{x}) = (\frac{h_c}{N})e^{-a\bar{x}}$ is assumed in the dry region ahead of wet front to simulate the effect of precursory cooling. Here, N, a, h_C represents the magnitude of precursory cooling, region of influence of precursory cooling and heat transfer coefficient in the wet region, respectively.

Earlier various analytical [15, 16, 17, 18] and numerical methods [19] have been employed to solve the conduction equation with precursory cooling. In this study an attempt has been made to obtain the solution of the rewetting model with precursory cooling by employing the variational integral method. The result obtained from the present analysis is compared with the available theoretical results and test data.

Theoretical analysis

Figure 4a schematically depicts the rewetting of a thin vertical slab of infinite length during top flooding. The assumptions common in the conduction-controlled rewetting model are presented in section 1.2.1. To simulate the effect of precursory cooling, the present model considers an exponentially varying heat transfer coefficient in the dry region ahead of wet front [10] (Figure4b). The transfer of energy from the hot surface to the surrounding fluid may be represented by a one dimensional control volume as shown in Figure4c. An elemental volume of length dx will be subjected to axial heat conduction. In addition, the hot wall transports q_v and q_c amount of energy to the surrounding fluid in the dry region and wet region, respectively. The heat loss from other side of the hot surface is considered to be negligible and assumed to be insulated. The one-dimensional transient heat conduction equation for the hot slab in Cartesian coordinate system, as shown in Figure4a can be expressed as:

$$K\frac{\partial^2 T}{\partial \bar{x}^2} - \frac{h}{\delta}(T - T_s) = \rho C\frac{\partial T}{\partial t}, \qquad (-\infty < \bar{x} < +\infty) \tag{37}$$

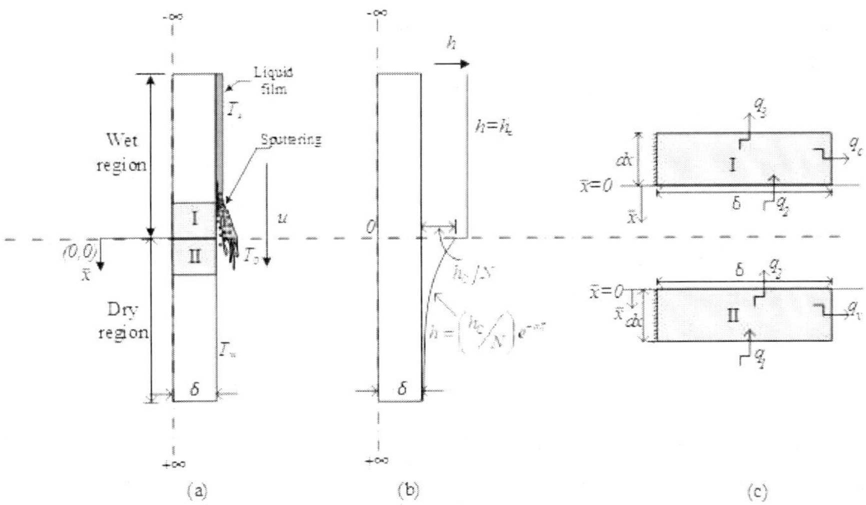

Figure 4. Schematic diagram of percursory cooling of one-dimensional hot object.

where, h is heat transfer coefficient. δ, ρ, C and K are thickness, density, specific heat, thermal conductivity of the slab material, respectively. Using quasi-steady state assumption ($\frac{\partial T}{\partial t} = -u\frac{\partial T}{\partial \bar{x}}$), Eq. (37) yields:

$$K\frac{\partial^2 T}{\partial \bar{x}^2} + \rho C u\frac{\partial T}{\partial \bar{x}} - \frac{h}{\delta}(T - T_s) = 0 \tag{38}$$

The following normalized variables are defined:

$$x = \frac{\bar{x}}{\delta},\ Pe = \frac{u\delta\rho C}{K},\ Bi = \frac{h\delta}{K},\ \theta = \frac{T - T_s}{T_0 - T_s},\ \theta_1 = \frac{T_w - T_0}{T_0 - T_s},\ \alpha = a\delta \tag{39}$$

Incorporating the effect of precursory cooling and utilizing (39), the energy equation (38) is transformed into following form:

$$\frac{d^2\theta}{dx^2} + Pe\frac{d\theta}{dx} - mBi\theta = 0 \begin{cases} m = 1; & wetregion \\ m = \frac{e^{-\alpha x}}{N}; & dryregion \end{cases} \tag{40}$$

Subjected to boundary condition given by:

$$B.C. = \begin{cases} \theta(-\infty) = 0, \theta(0) = 1; & (-\infty < x \leq 0) \\ \theta(0) = 1, \theta(+\infty) = 1 + \theta_1, & (0 \leq x \leq +\infty) \end{cases} \tag{41}$$

Solution Procedure

In the present study, the variational principle proposed by Arpaci [3] has been employed to solve the rewetting model. Accordingly, the governing differential equation of the physical problem is to be considered as the functional $F = f(x, \theta, \theta', \theta")$ for the analysis. Following Arpaci [3], the variational integral (I) valid for the present configuration can be expressed as:

$$I = \begin{cases} \int\limits_{-\infty}^{0} F dx, & -\infty < x \leq 0 \\ \int\limits_{0}^{+\infty} F dx, & 0 \leq x < +\infty. \end{cases} \tag{42}$$

Utilizing Eq. (40), (42) can be expressed as:

$$I = \begin{cases} \int\limits_{-\infty}^{0} \left(\frac{d^2\theta}{dx^2} + Pe\frac{d\theta}{dx} - Bi\theta\right) dx, & -\infty < x \leq 0 \\ \int\limits_{0}^{+\infty} \left(\frac{d^2\theta}{dx^2} + Pe\frac{d\theta}{dx} + Bi\frac{e^{-\alpha x}}{N}\theta\right) dx, & 0 \leq x < +\infty \end{cases} \tag{43}$$

At this juncture, it is necessary to assume a guess temperature profile that satisfies the boundary condition of the problem. There is hardly any guideline to select the best profile. Recently, Agrawal and Sahu [6, 20] have used a hybrid profile consists of exponential and polynomial functions of various orders to analyze the rewetting model. In the present configuration, the temperature decays (or increases slowly) away from the origin. Therefore, a hybrid profile involving both polynomial and exponential function is considered for the analysis and expressed as below.

$$\theta(x) = \begin{cases} P_1 + P_2 e^{x}(1 + a_1 x), & -\infty < x \leq 0 \\ P_3(1 - e^{-x}) + e^{-x}(1 + b_1 x), & 0 \leq x < +\infty \end{cases} \tag{44}$$

where P, a_1, b_1 are positive constants. To find the values of constants a_1, b_1 that will make the integral (I) a minimum, Eq. (43) is differentiated with respect to each constant and set the resulting equation to zero. Which yields:

$$\frac{\partial I}{\partial Y} = 0 \tag{45}$$

where, $Y = a_1, b_1$
Using boundary condition given in Eq. (41), and Eq. (43-45) one gets:

$$\theta(x) = \begin{cases} e^{x}(1 + a_1 x), & -\infty < x \leq 0 \\ (1 + \theta_1) - e^{-x}(\theta_1 - b_1 x), & 0 \leq x < +\infty \end{cases} \tag{46}$$

where,

$$a_1 = -\frac{1 + Pe - Bi}{1 + Bi}, b_1 = \frac{\frac{\theta_1(Pe-1)}{4} + \frac{Bi}{N}\left[\frac{\theta_1}{(2+\alpha)^2} - \frac{1+\theta_1}{(1+\alpha)^2}\right]}{\frac{2Bi}{N(2+\alpha)^3} + \frac{1}{4}} \tag{47}$$

It may be noted that the wet front velocity can be calculated by applying the continuity of heat flux at the interface as:

$$\left(\frac{d\theta}{dx}\right)_{x=o^+} = \left(\frac{d\theta}{dx}\right)_{x=o^-} \tag{48}$$

Which yields,

$$Pe = \frac{\left(\frac{2Bi}{1+Bi}\right)\left(\frac{1}{4} + \frac{2Bi}{N(2+\alpha)^3} - \frac{Bi}{N}\left[\frac{\theta_1}{(2+\alpha)^2} - \frac{1+\theta_1}{(1+\alpha)^2} + \frac{2\theta_1}{(2+\alpha)^3}\right]\right.}{\frac{\theta_1}{4} + \left(\frac{1}{1+Bi}\right)\left(\frac{1}{4} + \frac{2Bi}{N(2+\alpha)^3}\right)} \tag{49}$$

1.2.3. The Effect of Property Variation on Rewetting Velocity

It is observed that most of the rewetting models consider constant thermo-physical properties during analysis. It may be noted that during cooling of hot surfaces, metallurgical quenching and rewetting of nuclear fuel pin, the hot object undergoes a substantial change in the surface temperature. In such a case, assuming the constant thermo-physical property can only provide approximate results.

In view of this, it is necessary to include the effect of property variation in the rewetting analysis. Earlier, Olek and Zivrin [21] proposed a rewetting model that considers the effect of various temperature dependent properties such as: thermal conductivity, specific heat capacity, density and thickness of material for the analysis. Compared to other parameters, the variation of thermal conductivity with temperature was found to be more significant. Later on, Sahu et al.[22] considered the effect of temperature dependent thermal conductivity in their rewetting model. The Heat Balance Integral Method (HBIM) and optimal linearization technique was employed to obtain the solution.

In this study, the effect of temperature dependent thermal conductivity is considered in the basic rewetting model. The solution is obtained by employing the variational integral method. The result obtained by the present analysis is compared with other available results.

Theoretical analysis

The one-dimensional heat conduction equation for the hot solid in Cartesian coordinate system is expressed as:

$$\frac{\partial}{\partial \bar{x}}\left(K\frac{\partial T}{\partial \bar{x}}\right) - \frac{h}{\delta}(T - T_s) = \rho C\frac{\partial T}{\partial t}, \qquad (-\infty < \bar{x} < +\infty) \qquad (50)$$

where, h is heat transfer coefficient. δ, ρ, C and K are thickness, density, specific heat, thermal conductivity of the slab material, respectively. Using quasi-steady state assumption ($\frac{\partial T}{\partial t} = -u\frac{\partial T}{\partial \bar{x}}$), Eq. (50) yields:

$$\frac{\partial}{\partial \bar{x}}\left(K\frac{\partial T}{\partial \bar{x}}\right) - \frac{h}{\delta}(T - T_s) + \rho C u\frac{\partial T}{\partial \bar{x}} = 0, \qquad (-\infty < \bar{x} < +\infty) \qquad (51)$$

The variation of thermal conductivity of the hot solid is assumed to be a linear function of temperature and expressed as:

$$K = K_0\big[1 + \gamma(T - T_s)\big] = K_0\big[1 + \varepsilon_0\theta\big] \qquad (-\infty < \bar{x} < +\infty) \qquad (52)$$

Where K_0 is the thermal conductivity of the slab material at the ambient fluid temperature, ε_0 is the parameter which governs variation in thermal conductivity, and γ is constant such that $\varepsilon_0 << 1$. Following normalized parameters are defined:

$$x = \frac{\bar{x}}{\delta}, \quad Pe = \frac{u\delta\rho C}{K_0}, \quad Bi = \frac{h\delta}{K_0}, \quad \theta = \frac{T - T_s}{T_0 - T_s}, \quad \theta_1 = \frac{T_w - T_0}{T_0 - T_s}, \quad \varepsilon_0 = \gamma(T_0 - T_s) \quad (53)$$

where Pe is Peclet number, Bi is Biot number, T_0 is the quench front temperature corresponds to the temperature at minimum film boiling heat flux. θ and θ_1 are non-dimensional surface and dry wall temperature, respectively. Utilizing Eqs. (52-53), governing Eq. (51) reduces to:

$$\left[1 + \varepsilon_0\theta\right]\frac{d^2\theta}{dx^2} + \varepsilon_0\left(\frac{d\theta}{dx}\right)^2 + Pe\frac{d\theta}{dx} - mBi\theta = 0 \quad (54)$$

where,
$m = 1; \quad wetregion \quad (-\infty < x \leq 0)$
$m = 0; \quad dryregion \quad (0 \leq x < +\infty)$
Subjected to boundary condition given by:

$$B.C. = \begin{cases} \theta(-\infty) = 0, & \theta(0) = 1, \quad (-\infty < x \leq 0) \\ \theta(0) = 1, & \theta(+\infty) = 1 + \theta_1, \quad (0 \leq x < +\infty) \end{cases} \quad (55)$$

Solution procedure

In this section, we have employed variational approach as explained by Arpaci [3], the governing differential equation is considered as the functional $F = f(x, \theta, \theta', \theta")$ for the analysis. Following Arpaci [3], the variational integral (I) valid for the present configuration can be expressed as:

$$I = \begin{cases} \int_{-\infty}^{0} F dx, & -\infty < x \leq 0 \\ \int_{0}^{+\infty} F dx, & 0 \leq x < +\infty \end{cases} \quad (56)$$

Utilizing Eq. (54), Eq. (56) can be expressed as:

$$I = \begin{cases} \int_{-\infty}^{0} \left(\left[1 + \varepsilon_0\theta\right]\frac{d^2\theta}{dx^2} + \varepsilon_0\left(\frac{d\theta}{dx}\right)^2 + Pe\frac{d\theta}{dx} - Bi\theta\right)dx, & -\infty < x \leq 0 \\ \int_{0}^{+\infty} \left(\left[1 + \varepsilon_0\theta\right]\frac{d^2\theta}{dx^2} + \varepsilon_0\left(\frac{d\theta}{dx}\right)^2 + Pe\frac{d\theta}{dx}\right)dx, & 0 \leq x < +\infty \end{cases} \quad (57)$$

It may be noted that the accuracy of the solution obtained by employing the variational principle depends on the assumed guess profile. There is hardly any guideline to select the best profile. In the present configuration, the temperature decays (or increases slowly) away from the origin. Therefore, a hybrid profile involving both polynomial and exponential function is considered for the analysis and expressed as below.

$$\theta(x) = \begin{cases} P_1 + P_2 e^{x}(1 + a_1 x), & -\infty < x \leq 0 \\ P_3(1 - e^{-x}) + e^{-x}(1 + b_1 x), & 0 \leq x < +\infty \end{cases} \tag{58}$$

where $P, i = 1to3, a_1, b_1$ are positive constants. At this juncture, the variational integral (I) is minimized to evaluate the constants and is expressed as:

$$\frac{\partial I}{\partial Y} = 0 \tag{59}$$

where, $Y = a_1, b_1$
Using boundary condition given in Eq. (55), and Eq. (57-59) yields:

$$\theta(x) = \begin{cases} e^{x}(1 + a_1 x), & -\infty < x \leq 0 \\ (1 + \theta_1) - e^{-x}(\theta_1 - b_1 x), & 0 \leq x < +\infty \end{cases} \tag{60}$$

Eq. (60) represents the distribution of temperature in the hot solid for both wet and dry region. It may be noted that the wet front velocity can be calculated by applying the continuity of heat flux at the interface as:

$$\left(\frac{d\theta}{dx}\right)_{x=o^+} = \left(\frac{d\theta}{dx}\right)_{x=o^-} \tag{61}$$

Which yields,

$$1 + a_1 = \theta_1 + b_1 \tag{62}$$

Where
constants=

$$\begin{cases} a_1 = \left\{ \dfrac{16\varepsilon_0 + 27(1 + Bi) - \sqrt{640(\varepsilon_0)^2 + 864\varepsilon_0(1 + Bi) + 729(1 + Bi)^2 + 432\varepsilon_0(1 + Pe - Bi)}}{8\varepsilon_0} \right. \\ b_1 = \left\{ \dfrac{27(1 + \varepsilon_0) - 21\varepsilon_0\theta_1 - \sqrt{(21\varepsilon_0\theta_1 - 27\varepsilon_0 - 27)^2 - 352\varepsilon_0(27Pe\theta_1 - 27Pe - 3\varepsilon_0\theta_1^2 - 27\varepsilon_0\theta_1)}}{176\varepsilon_0} \right. \end{cases} \tag{63}$$

1.2.4. Multiregion Rewetting Model [23]

The two-region rewetting model usually assumes a constant heat transfer coefficient in the wet region and an adiabatic condition in the dry region ahead of wet front. Experimental observations reveal that the heat removal rate is significant over a narrow region near the quench front. The higher heat removal rate may be due to the turbulent nature of the bubbles and higher boiling heat transfer coefficient. This narrow region is termed as sputtering region and found to influence strongly the rewetting velocity. Therefore, a single heat transfer coefficient in the entire wet region may not be suitable for analyzing the phenomenon of rewetting. In order to model the sputtering region, different heat transfer coefficients are assumed in the wet region.

The rewetting analysis of hot surface for several distributions of heat transfer coefficient along the hot surface is obtained by employing variational integral method. Closed form solution is obtained for the temperature field and rewetting velocity. Present variational integral method solution incorporates the sputtering region in the model. By neglecting the sputtering region, the three-region model reduces to the two-region model. The results of the present analysis are compared with other available analytical results and test data.

Theoretical analysis The physical model under consideration is an infinitely extended vertical slab with the liquid supplied from the top as shown in Figure 5a. As the liquid film progresses downward, vapour is generated near the liquid front and a thin vapour film is formed at the solid-liquid interface that prevents the liquid from contacting the hot surface. As the process continues, the surface temperature cools off from its initial wall temperature T_w to rewetting temperature T_0, at which the vapor blanket becomes unstable and collapses. Consequently, the liquid wets the hot surface. The region behind the wet front corresponds to transition boiling and is followed by nucleate boiling regime. Beyond this, the surface temperature drops below the temperature of incipient boiling and the heat is removed by convection to single-phase liquid. Following assumptions are made for the analysis [23]:

1. The wet front velocity is assumed to be constant during its propagation. The end effects are neglected; therefore the problem can be reduced to a quasi-steady one [8, 24].

2. In the present one dimensional model, the hot solid is divided into three distinct regions, dry region ahead of the wet front ($0 \leq \bar{x} < +\infty$), the

sputtering region immediately behind the wet front $(-l \leq \bar{x} \leq 0)$ and a continuous film region further upstream $(-\infty < \bar{x} \leq -l)$. In the sputtering region $(-l \leq \bar{x} \leq 0)$, the temperature of the hot surface varies from rewetting temperature T_0 to incipient boiling temperature T_b and in the liquid region $(-\infty < \bar{x} \leq -l)$ the surface temperature varies from T_b to liquid saturation temperature T_s.

3. Two constant but different heat transfer coefficients namely, h_C and h_B are considered for the wet and the sputtering regions, respectively. However, for dry region, an adiabatic condition $(h = 0)$ is considered [5, 26] (Figure 5b).

4. Suitable values of rewetting temperature and heat transfer coefficients are taken as input parameters to solve the conduction equation.

5. Other factors such as: surface finish, surface roughness, etc. are either neglected or they do not affect the process of rewetting.

The one-dimensional conduction equation for the hot solid in Cartesian coordinate system (Figure 5) can be written as:

$$K\frac{\partial^2 T}{\partial \bar{x}^2} - \frac{h}{\delta}(T - T_s) = \rho C\frac{\partial T}{\partial t}, \qquad (-\infty < \bar{x} < +\infty) \tag{64}$$

where K, ρ, C and δ represents thermal conductivity, density, specific heat capacity and thickness of the slab material, respectively. While, h represents the wet side heat transfer coefficient. Using the quasi-steady state assumption $(\frac{\partial T}{\partial t} = -u\frac{\partial T}{\partial \bar{x}})$, Eq. (64) yields:

$$K\frac{\partial^2 T}{\partial \bar{x}^2} + \rho C u\frac{\partial T}{\partial \bar{x}} - \frac{h}{\delta}(T - T_s) = 0 \qquad (-\infty < \bar{x} < +\infty) \tag{65}$$

The following normalized variables are defined:

$$\begin{aligned} x &= \frac{\bar{x}}{\delta}, \; Pe = \frac{u\delta\rho C}{K}, \; Bi = \frac{h\delta}{K}, \; \theta = \frac{T - T_s}{T_0 - T_s}, \theta_1 = \frac{T_w - T_0}{T_0 - T_s}, \\ \theta_2 &= \frac{T_b - T_s}{T_0 - T_s}, \quad L = \frac{l}{\delta}, \quad Bi_B = \frac{h_B\delta}{K}, \quad Bi_C = \frac{h_C\delta}{K} \end{aligned} \tag{66}$$

Utilizing Eq. (66), the energy equation (65) is transformed into following form:

$$\frac{d^2\theta}{dx^2} + Pe\frac{d\theta}{dx} - J\theta = 0 \tag{67}$$

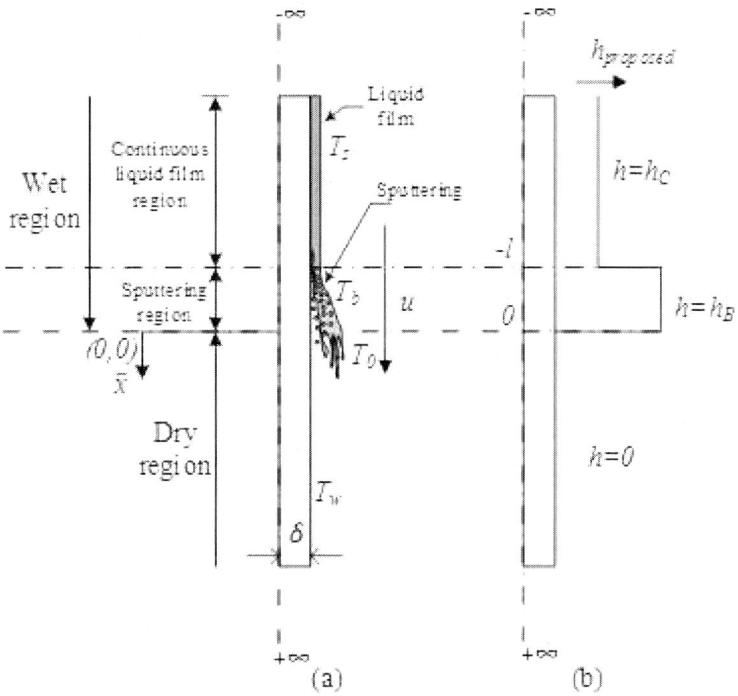

Figure 5. Schematic of one-dimensional hot object [23].

where,

$$J = \begin{cases} Bi_C; & (-\infty < x < -L) \\ Bi_B; & (-L < x \leq 0) \\ 0; & (0 \leq x < \infty) \end{cases} \tag{68}$$

Subjected to boundary conditions given by:

$$\theta(-\infty) = 0, \qquad \theta(-L) = \theta_2 \qquad \theta(0) = 1, \qquad \theta(+\infty) = 1 + \theta_1; \tag{69}$$

Solution procedure

In this section, the variational principle proposed by Arpaci [3] has been employed to solve the conduction equation. In such a case, the governing differential equation of the physical problem is considered as the functional

$F = f(x, \theta, \theta', \theta'')$ for the analysis. Following Arpaci [3], the variational integral (I) valid for the present configuration can be expressed as:

$$I = \begin{cases} \int_{-\infty}^{-L} F dx, & -\infty < x \leq -L \\ \int_{-L}^{0} F dx, & -L \leq x \leq 0 \\ \int_{0}^{+\infty} F dx, & 0 \leq x < +\infty \end{cases} \tag{70}$$

Utilizing Eq. (67), Eq. (70) can be expressed as:

$$I = \begin{cases} \int_{-\infty}^{-L} \left(\frac{d^2\theta}{dx^2} + Pe \frac{d\theta}{dx} - Bi_C \theta \right) dx, & -\infty < x \leq -L \\ \int_{-L}^{0} \left(\frac{d^2\theta}{dx^2} + Pe \frac{d\theta}{dx} - Bi_B \theta \right) dx, & -L \leq x \leq 0. \\ \int_{0}^{+\infty} \left(\frac{d^2\theta}{dx^2} + Pe \frac{d\theta}{dx} \right) dx, & 0 \leq x < +\infty \end{cases} \tag{71}$$

At this juncture, it is necessary to assume a guess temperature profile that satisfies the boundary condition of the physical problem. Although the accuracy of the solution obtained by the variational integral method depends on the guess profile, there is hardly any guideline to select the best profile. Previously, Agrawal and Sahu [6, 20] used the hybrid profile involving the exponential and polynomial functions of different orders to analyze the basic rewetting model. It is observed that the temperature decays (or increase) exponentially farthest from the quench front in the wet region or dry region respectively. In view of this, a hybrid profile involving both polynomial and exponential function is considered in the wet region and the dry region. While three different guess profiles involving various orders of polynomial functions are tried in the sputtering region. The details of the various models are elaborated below.

Model (1)

This model considers a hybrid profile involving both polynomial and exponential function in the wet and dry region; polynomial guess profile in the sputtering region as below.

$$\theta(x) = \begin{cases} e^{x+L}(\theta_2 + a_1(x+L)), & -\infty < x \leq -L \quad (72) \\ a_2x^2 + a_3x + a_4, & -L \leq x \leq 0. \quad (73) \\ (1+\theta_1)(1-e^{-x}) + e^{-x}(1+a_5x), & 0 \leq x < +\infty \quad (74) \end{cases}$$

where $a_1 ... a_5$ are positive constants. At this juncture, the variational integral (I) is minimized to evaluate the constants and is expressed as:

$$\frac{\partial I}{\partial Y} = 0 \tag{75}$$

where, $Y = a_1, b_1$
and,

$$\begin{aligned} a_1 &= -\frac{\theta_2(1+Pe-Bi_C)}{1+Bi_C}, \\ a_2 &= -\frac{10Pe(1-\theta_2)+5Bi_BL(1-\theta_2)-10Bi_BL}{2Bi_BL^3+20L}, \quad a_4 = 1, \quad a_5 = \theta_1(Pe-1), \\ a_3 &= La_2 + \frac{(1-\theta_2)}{L}, \end{aligned} \tag{76}$$

It may be noted that utilizing Eq. (76) and Eqs. (72-74), one can evaluate the temperature distribution in various regions, namely, wet, sputtering and dry region the hot object. While, the non-dimensional wet front velocity, Pe, and the dimensionless sputtering length, L, are still unknown and they are determined by using the continuity of heat flux at the interface as below.

$$\left(\frac{d\theta}{dx}\right)_{x=-L^+} = \left(\frac{d\theta}{dx}\right)_{x=-L^-} and \left(\frac{d\theta}{dx}\right)_{x=o^+} = \left(\frac{d\theta}{dx}\right)_{x=o^-} \tag{77}$$

Combining Eqs. (72-74), (76) with Eq.(77), the values Pe and L are determined by the simultaneous equations expressed as:

$$Pe\left(\frac{10(1-\theta_2)}{2Bi_BL^2+20} + \frac{\theta_2}{1+Bi_C}\right) = \frac{2Bi_C\theta_2}{1+Bi_C} + \frac{5Bi_BL(1+\theta_2)}{2Bi_BL^2+20} + \frac{\theta_2-1}{L} \tag{78}$$

and

$$Pe\left(\frac{10(1-\theta_2)}{2Bi_BL^2+20} + \theta_1\right) = \frac{5Bi_BL(1+\theta_2)}{2Bi_BL^2+20} + \frac{1-\theta_2}{L} \tag{79}$$

Eqs. (78) and (79) present the relationship between various modeling parameters, namely, $Pe, L, Bi_B, Bi_C, \theta_1$ and θ_2. For known values of Bi_B, Bi_C, θ_1 and θ_2, one can evaluate the Peclet number (Pe) and sputtering length (L) by utilizing a suitable iterative technique. Subsequently, the temperature profile for various regimes can be evaluated by employing Eq. (76).

Model (2)

The present configuration utilizes the hybrid guess profile expressed in Eq. (72) and Eq. (74) for the wet region and dry region, respectively. However, in the case of sputtering region, a cubic polynomial function is considered as below.

$$\theta(x) = b_1 x^3 + b_2 x + b_3, \qquad -L \leq x \leq 0 \tag{80}$$

Employing Eq. (75), constants $b_1 ... b_3$ can be expressed as:

$$b_1 = -\frac{7}{4}\left(\frac{-15Pe(1-\theta_2) + (7+8\theta_2)Bi_B L}{84L^2 + 8Bi_B L^4}\right), \quad b_2 = \frac{1 - b_1 L^3 - \theta_2}{L}, \quad b_3 = 1 \tag{81}$$

Combining Eqs. (76), (81) and Eqs. (72), (74) and (80), one can evaluate the temperature distribution in various regions (wet, sputtering and dry) of the hot object. While, the non-dimensional wet front velocity (Pe) and the dimensionless sputtering length (L), can be evaluated utilizing Eq. (77). These yields:

$$Pe\left(\frac{105(1-\theta_2)}{2(84 + 8Bi_B L^2)} + \frac{\theta_2}{1 + Bi_C}\right) = \frac{2Bi_C\theta_2}{1 + Bi_C} + \frac{7}{2}\frac{(7+8\theta_2)Bi_B L}{84 + 8Bi_B L^2} + \frac{\theta_2 - 1}{L} \tag{82}$$

and

$$Pe\left(\theta_1 + \frac{105(1-\theta_2)}{2(84 + 8Bi_B L^2)}\right) = \frac{1-\theta_2}{L} + \frac{7}{4}\left(\frac{(7+8\theta_2)Bi_B L}{84 + 8Bi_B L^2}\right) \tag{83}$$

Using the Eqs. (82) and (83) and by employing a suitable iterative technique, one can evaluate the Peclet number (Pe) and sputtering length (L). Therefore, the temperature distribution along the hot object can be evaluated by employing Eqs. (80), (81), (72), (74) and (76).

Model (3)

This model utilizes the hybrid guess profile expressed in Eqs.(72) and (74) for the wet region and dry region, respectively. However, in the case of sputtering region, a cubic polynomial function is considered as below.

$$\theta(x) = c_1 + c_2(2Lx - x^2) + c_3(3Lx^2 - 2x^3), \qquad -L \leq x \leq 0 \tag{84}$$

Following Eq. (75), constants $c_1 ... c_3$ and Peclet number can be evaluated and expressed as:

$$c_1 = 1, \; c_2 = -\frac{42(133 - 55PeL)(\theta_2 - 1) - (1459 + 641\theta_2)Bi_B L^2}{3052L^2 + 298Bi_B L^4}, \; c_3 = \frac{\theta_2 - 1 + 3L^2 c_2}{5L^3} \tag{85}$$

Combining Eqs. (76), (84), (85) with Eqs. (72), (74), one can evaluate the temperature distribution in various regions (wet, sputtering and dry) of the hot object. Utilizing Eq. (77), one can obtain:
Pe=

$$\frac{\left(\frac{872L^2}{375} + \frac{596Bi_B L^4}{2625}\right)\left(\frac{10Bi_C\theta_2}{16L(1+Bi_C)} + \frac{3(\theta_2-1)}{4L^2}\right) - \left(\frac{133(\theta_2-1)}{125}\right) + \left(\frac{(1459+641\theta_2)Bi_B L^2}{5250}\right) + \left(\frac{55L(\theta_2-1)}{125}\right)}{\left(\frac{872L^2}{375} + \frac{596Bi_B L^4}{2625}\right)\left(\frac{5\theta_2}{16L(1+Bi_C)}\right)} \tag{86}$$

and,

$$Pe = \frac{\left(\frac{133(\theta_2-1)}{125}\right) - \left(\frac{(1459+641\theta_2)Bi_B L^2}{5250}\right) + \left(\frac{872L^2}{375} + \frac{596Bi_B L^4}{2625}\right)\left(\frac{\theta_1}{2L}\right)}{\left(\frac{55L(\theta_2-1)}{125}\right)} \tag{87}$$

Using the equations (86) and (87) and by employing a suitable iterative technique, one can evaluate the Peclet number (Pe) and sputtering length (L). Therefore, the temperature distribution along the hot object can be evaluated by employing Eqs. (72), (74), and (76), (84) and (85).

1.2.5. Multi-Region Rewetting Model with Precursory Cooling [25]

Several rewetting models have been proposed that consider two or three different regions for analysis. Usually a constant heat transfer coefficient of different magnitude is assumed in the wet region; while, an adiabatic condition is assumed in the dry region ahead of wet front. In actual situation, the distribution of heat transfer coefficient near the sputtering region and the dry region ahead of the wet front varies along the axial direction [27]. It is argued that accurate choice of shape of the heat transfer coefficient is important to precisely predict the wet front velocity and surface temperature distribution. In this study an attempt has been made to extend the previous work (section 1.2.1) to analyze the multi-region rewetting model with varying heat transfer coefficient in the sputtering and dry region ahead of the wet front. Two different rewetting models are considered for the analysis. In the first model, a constant heat transfer coefficient is considered in the sputtering region and in second model, an axially varying heat transfer coefficient in the sputtering region is considered. Both the models consider an exponentially decaying heat transfer coefficient in the dry region ahead of wet front. For all the cases the closed form expression is obtained for temperature field along axial direction. Relationship between various rewetting parameters such as: Peclet number, Biot number, dry wall temperature, incipient boiling temperature, sputtering length, magnitude of precursory cooling and the extent of precursory cooling have been obtained from the analysis.

Theoretical analysis

Figure 6a schematically depicts the falling film rewetting of a one-dimensional slab of infinite length. As described in the previous section (section 1.2.4), three different boiling regimes are observed behind the wet front during its propagation. In addition to this, the dry wall ahead of wet front is cooled due to precursory cooling (section 1.2.2). In such a case the distribution of heat transfer along the hot object becomes non-linear and the heat transfer profile varies arbitrarily along the axial direction (Figure 6b).

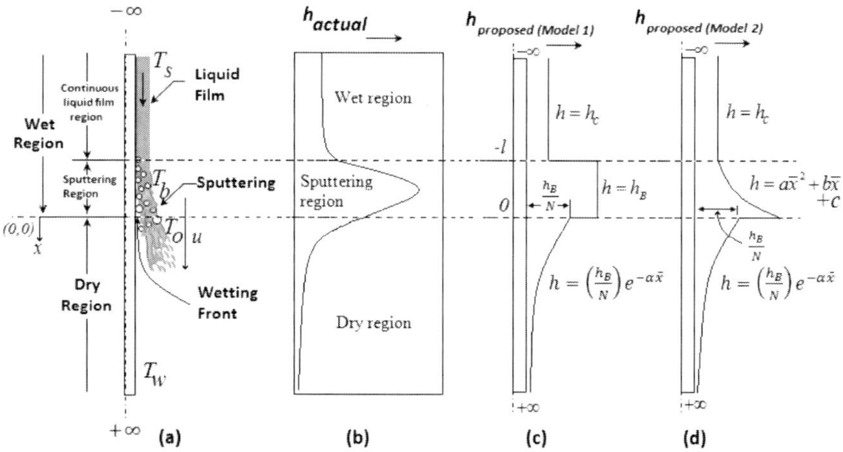

Figure 6. Schematic of three-region model of one-dimensional hot object [25].

Figure 6b depicts the actual variation of heat transfer coefficient along axial direction. It is shown that the magnitude of heat transfer coefficient near the wet front location is very high and it gradually decreases in the downstream direction.

In order to analyze the rewetting phenomenon one can assume a constant heat transfer coefficient of different magnitude in both wet and sputtering region. Alternatively, one can consider a constant heat transfer coefficient in the wet region and a parabolic variation in heat transfer coefficient in the sputtering region. The assumptions applicable for conduction controlled rewetting are reported in section 1.2.4(a, b, d,and e). In addition, present model considers a constant heat transfer coefficient (h_C) in the wet region and an exponentially decaying heat transfer coefficient of the form $h(\bar{x}) = (\frac{h_B}{N})e^{-a\bar{x}}$ is considered in the dry region ahead of wet front [10]. Two different models are considered to account the variation in heat transfer in the sputtering region. First model considers a constant heat transfer coefficient (h_B) in the sputtering region (Figure6c). While, the other model considers, an axially varying heat transfer coefficient of the form $h(\bar{x}) = b\bar{x}^2 + c\bar{x} + d'$ in the sputtering region (Figure6d). It may be noted that for a given value of Biot number and dry wall temperature, one-dimensional model predicts lower rewetting velocity compared to the two-dimensional model. In order to avoid complexity of numerical calculation, a one-dimensional model is proposed to demonstrate the effect of precursory cooling and variation in heat transfer coefficient in the sputtering region. The transfer of energy from the hot surface to the surrounding fluid may be represented by a one-dimensional control volume as shown in Figure7. An elemental volume of length dx will be subjected to axial heat conduction. In addition the hot wall transports q_v, q_b and q_c amount of energy to the surrounding fluid in the dry, sputtering and wet region, respectively. The heat loss from other side of the hot surface is considered to be negligible and assumed to be insulated.

The one-dimensional conduction equation for the hot solid in Cartesian coordinate system (Figure 6a) can be written as:

$$K\frac{\partial^2 T}{\partial \bar{x}^2} - \frac{h}{\delta}(T - T_s) = \rho C\frac{\partial T}{\partial t}, \qquad (-\infty < \bar{x} < +\infty) \tag{88}$$

where K, ρ, C and δ represents thermal conductivity, density, specific heat capacity and thickness of the slab material, respectively. While, h represents the wet side heat transfer coefficient. Using the quasi-steady state assumption $(\frac{\partial T}{\partial t} = -u\frac{\partial T}{\partial \bar{x}})$, Eq. (88) yields:

$$K\frac{\partial^2 T}{\partial \bar{x}^2} + \rho C u\frac{\partial T}{\partial \bar{x}} - \frac{h}{\delta}(T - T_s) = 0 \qquad (-\infty < \bar{x} < +\infty) \tag{89}$$

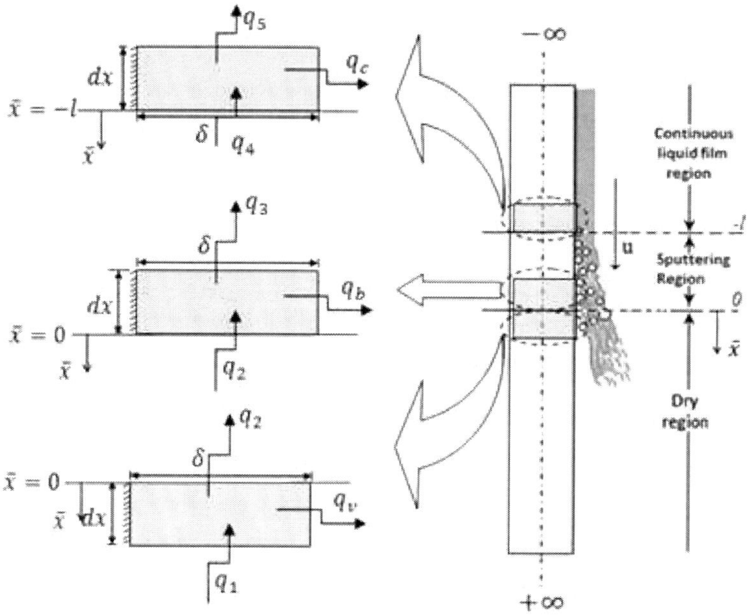

Figure 7. Control volume analysis of one-dimensional hot surface [25].

The following normalized variables are defined:

$$\begin{aligned} x &= \frac{\bar{x}}{\delta}, \; Pe = \frac{u\delta\rho C}{K}, \; \theta = \frac{T - T_s}{T_0 - T_s}, \; \theta_1 = \frac{T_w - T_0}{T_0 - T_s}, \; \theta_2 = \frac{T_b - T_s}{T_0 - T_s}, \\ L &= \frac{l}{\delta}, \quad Bi_B = \frac{h_B\delta}{K}, \quad Bi_C = \frac{h_C\delta}{K}, \alpha = a\delta \end{aligned} \tag{90}$$

Utilizing Eq. (90), the energy equation (89) is transformed into following form:

$$\frac{d^2\theta}{dx^2} + Pe\frac{d\theta}{dx} - J\theta = 0 \tag{91}$$

Subjected to boundary conditions given by:

$$\theta(-\infty) = 0, \qquad \theta(-L) = \theta_2 \qquad \theta(0) = 1, \qquad \theta(+\infty) = 1 + \theta_1; \tag{92}$$

The value of J varies for different heat transfer regimes and the details of the heat transfer models are elaborated below.

Model (1):Constant heat transfer coefficient (h_B) *in the sputtering region*

The model includes two constant and different heat transfer coefficients, namely, h_C and h_B in the wet and the sputtering region, respectively. While, the dry region consider an exponentially decaying heat transfer coefficient of the form $h(\bar{x}) = (\frac{h_B}{N})e^{-a\bar{x}}$ (Figure 6c). In such a case the value of J in Eq. (91) is expressed as:

$$J = \begin{cases} Bi_C, & -\infty < x \leq -L \\ Bi_B, & -L \leq x \leq 0. \\ \frac{e^{-\alpha x}}{N} Bi_B, & 0 \leq x < +\infty \end{cases} \tag{93}$$

In the present study, the variational principle proposed by Arpaci [3] has been employed to solve the conduction equation. The governing differential equation of the physical problem (91) is considered as Eulers equation of the variational formulation. Following Arpaci [3], the variational integral (I) valid for the present configuration can be expressed as:

$$I = \begin{cases} \int\limits_{-\infty}^{-L} F dx, & -\infty < x \leq -L \\ \int\limits_{-L}^{0} F dx, & -L \leq x \leq 0 \\ \int\limits_{0}^{+\infty} F dx, & 0 \leq x < +\infty \end{cases} \tag{94}$$

Utilizing Eq. (91) and (93), Eq. (94) can be expressed as:

$$I = \begin{cases} \int\limits_{-\infty}^{-L} \left(\frac{d^2\theta}{dx^2} + Pe\frac{d\theta}{dx} - Bi_C\theta\right) dx, & -\infty < x \leq -L \\ \int\limits_{-L}^{0} \left(\frac{d^2\theta}{dx^2} + Pe\frac{d\theta}{dx} - Bi_B\theta\right) dx, & -L \leq x \leq 0. \\ \int\limits_{0}^{+\infty} \left(\frac{d^2\theta}{dx^2} + Pe\frac{d\theta}{dx} - Bi_B\frac{e^{-\alpha x}}{N}\theta\right) dx, & 0 \leq x < +\infty \end{cases} \tag{95}$$

At this juncture, it is necessary to assume a guess temperature profile that satisfies the boundary condition of the physical problem. Although the accuracy of the solution obtained by the variational integral method depends on the guess profile, there is hardly any guideline to select the best profile. Previously, Agrawal and Sahu [6, 20] used the hybrid profile involving the exponential and polynomial functions of different orders to analyze the basic rewetting model. In this study, a hybrid profile involving both polynomial and exponential function is considered in the wet region $(-\infty < x \leq -L)$ and the dry region $(0 \leq x < +\infty)$ respectively. While, parabolic polynomial function is tried in the sputtering region $(-L \leq x \leq 0)$ and is expressed as:

$$\theta(x) = \begin{cases} M_1 + Me^{x+L}(\theta_2 + a_1(x+L)), & -\infty < x \leq -L \quad (96) \\ \theta(x) = M_3 a_2 x^2 + M_4 a_3 x + M_5 a_4, & -L \leq x \leq 0 \quad (97) \\ \theta(x) = M_6(1+\theta_1)(1-e^{-x}) + e^{-x}(1+a_5 x), & 0 \leq x < +\infty \quad (98) \end{cases}$$

where $M_1 ... M_3$ and $a_1 ... a_5$ are positive constants.
It may be noted that the Eq. (96-98) satisfies the boundary conditions of the physical problem. In order to evaluate the constants one needs to differentiate the variational integral (I) with respect to each constant. This is expressed as:

$$\frac{\partial I}{\partial Y} = 0 \tag{99}$$

where, $y = a_1 ... a_5$
By using Eqs. (95-99) and boundary conditions Eq. (92)

$$\begin{cases} \theta(x) = e^{x+L}(\theta_2 + a_1(x+L)), & -\infty < x \leq -L \quad (100) \\ \theta(x) = a_2 x^2 + a_3 x + a_4, & -L \leq x \leq 0 \quad (101) \\ \theta(x) = (1+\theta_1)(1-e^{-x}) + e^{-x}(1+a_5 x), & 0 \leq x < +\infty \quad (102) \end{cases}$$

where,

$$a_1 = -\frac{\theta_2(1+Pe-Bi_C)}{1+Bi_C}, \quad a_4 = 1,$$
$$a_2^1 = -\frac{10Pe(1-\theta_2) + 5Bi_B L(1-\theta_2) - 10Bi_B L}{2Bi_B L^3 + 20L}, \quad a_5 = \frac{\frac{\theta_1(Pe-1)}{4} + \frac{Bi_B}{N}\left[\frac{\theta_1}{(2+\alpha)^2} - \frac{1+\theta_1}{(1+\alpha)^2}\right]}{\left[\frac{2Bi_B}{N(2+\alpha)^3} + \frac{1}{4}\right]}$$
$$a_3 = La_2^1 + \frac{(1-\theta_2)}{L}, \tag{103}$$

It may be noted that utilizing Eq. (103) and Eqs. (100-102), one can evaluate the temperature distribution in various regions, namely, wet, sputtering and dry region of the hot object. While, the non-dimensional wet front velocity, Pe, and the dimensionless sputtering length, L, are still unknown and they can be determined by using the continuity of heat flux at the interface as below.

$$\left(\frac{d\theta}{dx}\right)_{x=-L^+} = \left(\frac{d\theta}{dx}\right)_{x=-L^-} and \left(\frac{d\theta}{dx}\right)_{x=o^+} = \left(\frac{d\theta}{dx}\right)_{x=o^-} \tag{104}$$

Combining Eqs. (101-103) with Eq.(104), the values Pe and L are determined by the simultaneous equations expressed as:

$$Pe\left(\frac{10(1-\theta_2)}{2Bi_B L^2+20} + \frac{\theta_2}{1+Bi_C}\right) = \frac{2Bi_C\theta_2}{1+Bi_C} + \frac{5Bi_B L(1+\theta_2)}{2Bi_B L^2+20} + \frac{\theta_2 - 1}{L} \tag{105}$$

and

$$La_2^1 + \frac{1-\theta_2}{L} = \theta_1 + a_5 \tag{106}$$

Eq. (105) and (106) are of great interest as these represent the relationship among various modeling parameters, namely, Peclet number (Pe), sputtering length (L), boiling Biot number (Bi_B), convective Biot number (Bi_C), dry wall temperature (θ_1), sputtering temperature (θ_2), magnitude of precursory cooling (N) and the region for the influence of precursory cooling (α). For the case of no precursory cooling $(N \to \infty)$, the results of the present model Eqs. (105) and (106) are presented and compared with other analytical result.

Model (2):Axially varying heat transfer coefficient (h_B) *in the sputtering region*

The model considers a constant heat transfer coefficient (h_C) in the wet region, parabolic variation in the heat transfer coefficient in sputtering region and an exponentially decreasing heat transfer coefficient is considered in the dry region for the analysis (Figure 6d). In this case the value of J in Eq. (91) is expressed as:

$$J = \begin{cases} Bi_C, & -\infty < x \leq -L \\ bx^2 + cx + d', & -L \leq x \leq 0. \\ \frac{e^{-\alpha x}}{N} d', & 0 \leq x < +\infty \end{cases} \tag{107}$$

The profile $bx^2 + cx + d'$ involves various constants. Utilizing the boundary condition $(x = -L, Bi_B = Bi_C)$ and $(x = 0, \theta_2 = 1)$, the constants for the heat transfer profile are obtained as below.

$$b = \frac{Bi_B - Bi_C}{L^2}, \quad c = \frac{2(Bi_B - Bi_C)}{L}, \quad d' = Bi_B \tag{108}$$

The variational integral method proposed by Arpaci [3] is employed in order to solve the conduction equation. Following the procedure described in Model (1), and utilizing Eqs (91), (107), (108) and (94), one can express:

$$I = \begin{cases} \int_{-\infty}^{-L} \left(\frac{d^2\theta}{dx^2} + Pe\frac{d\theta}{dx} - Bi_C\theta\right)dx, & -\infty < x \leq -L \\ \int_{-L}^{0} \left(\frac{d^2\theta}{dx^2} + Pe\frac{d\theta}{dx} - \left(\frac{(Bi_B - Bi_C)}{L}x^2 + \frac{2(Bi_B - Bi_C)}{L}x + Bi_B\right)\theta\right)dx, & -L \leq x \leq 0. \\ \int_{0}^{+\infty} \left(\frac{d^2\theta}{dx^2} + Pe\frac{d\theta}{dx} - Bi_B\frac{e^{-\alpha x}}{N}\theta\right)dx, & 0 \leq x < +\infty \end{cases} \tag{109}$$

Utilizing Eq. (109), Eqs. (96-99) and boundary condition Eq. (92) one can obtain Eqs. (100-102), where

$$\begin{aligned} a_1 &= -\frac{\theta_2(1 + Pe - Bi_C)}{1 + Bi_C}, \\ a_2^2 &= -\frac{70Pe(1-\theta_2) - 7L(1-\theta_2)(3c - 2b - 5d') - 7L(3b - 5c + 10d')}{L^3(-4b + 7c - 14d') - 140L}, \\ a_3 &= La_2^2 + \frac{(1-\theta_2)}{L}, \\ a_4 &= 1, \\ a_5 &= \frac{\frac{\theta_1(Pe-1)}{4} + \frac{Bi_B}{N}\left[\frac{\theta_1}{(2+\alpha)^2} - \frac{1+\theta_1}{(1+\alpha)^2}\right]}{\left[\frac{2Bi_B}{N(2+\alpha)^3} + \frac{1}{4}\right]} \end{aligned} \tag{110}$$

Combining Eqs.(100-102) and Eq. (110), one can evaluate the temperature distribution in various regions (wet, sputtering and dry) of the hot object. While, the non-dimensional wet front velocity, Pe, and the dimensionless sputtering length, L, can be evaluated utilizing Eq. (104). These yields:

$$-Pe\left(\frac{70(1-\theta_2)}{Z} - \frac{\theta_2}{1 + Bi_C}\right) = \frac{2Bi_C\theta_2}{1 + Bi_C} - \frac{7L(1-\theta_2)(3c - 2b - 5d') + 7L(3b - 5c + 10d')}{Z} - \frac{1-\theta_2}{L} \tag{111}$$

where;
$Z = L^2(-4b + 7c - 14d') - 140$
and

$$a_2^2 L + \left(\frac{1-\theta_2}{L}\right) = \theta_1 + a_5 \tag{112}$$

Eqs. (111) and (112) represents the relationship among various modeling parameters, namely, Peclet number (Pe), sputtering length (L), boiling Biot number (Bi_B), convective Biot number (Bi_C), dry wall temperature (θ_1), sputtering temperature (θ_2), magnitude of precursory cooling (N) and the region for the influence of precursory cooling (α). Using the Eqs. (111) and (112) and by employing a suitable iterative technique, one can evaluate the Peclet number (Pe) and sputtering length (L). The present model (Eqs. 111-112) is correlated with the test data for a given set of rewetting parameters and is reported in the subsequent section.

2. Results and Discussion

The present chapter reports the application of variational integral method proposed by Sparrow and Siegel [2] and Arpaci [3] to analyze the rewetting behavior of a hot surface. Both these approaches use different functions to evaluate the variational integral. Subsequently, in both the cases, the variational integral is minimized to evaluate the constants used in the guess temperature profile. Based on the analysis a closed form expression is obtained between various parameters such as: Peclet number, Biot number and dry wall temperature. The results are discussed in the following section. A wide variety of theoretical models have been proposed to analyze the rewetting behavior of hot surface. These models employ either an analytical method or numerical one to solve the conduction equation. Most of the theoretical models usually predict the temperature distribution, rewetting temperature, and rewetting velocity during the phenomenon of rewetting. The theoretical models adopted the analytical techniques, namely, separation of variables, Winer-Hopf technique, and Heat Balance Integral Method and numerical techniques such as: finite difference technique and the finite element method. In this study, an attempt has been made to apply variational integral method to solve the conduction equation and obtained a three parametric variation between various parameters such as: dry wall temperature, Biot number and Peclet number. It may be noted that the results obtained by employing Sparrow approach [2] matches exactly with other theoretical model [4, 8, 24] and hence not used for the comparison. While the results obtained by employing Arpaci approach [3] with a second order approximation is used to compare with other theoretical models. Present prediction is compared with the results obtained by employing finite difference method [6], separation of variable method [8, 29] and Heat Balance Integral method [4] and is depicted in Figure 8. The agreement between present predictions obtained by a second order approximation with other model is excellent in certain cases. It may be noted that the agreement between various reported predictions is poor for certain values of Biot number. An order of magnitude difference among some of the predictions are noted and shown in Figure 8.

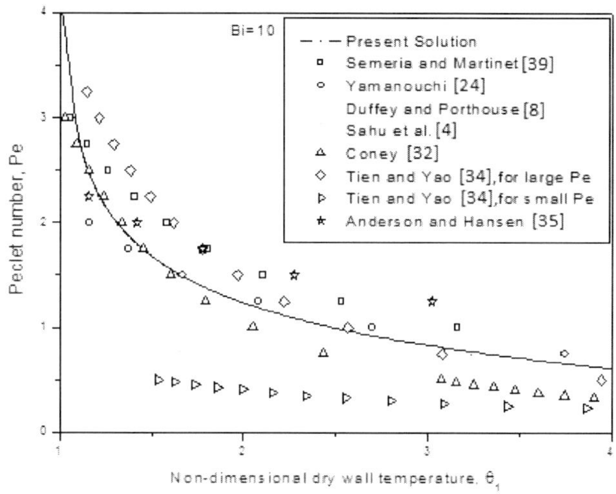

Figure 8. Comparison of VIM model with other analytical models [6].

It is evident from the literature that a host of experiments have been conducted in order to analyze the rewetting of hot surfaces in various geometries such as: slab, tube and rod [24, 30, 31] for a wide range of coolant flow rate. Tests have been carried out either in single rod [32] or with rod bundle [24, 30, 33] for top flooding. It is found that, most of the experiments were carried out using different coolants for atmospheric pressure condition [24, 30, 33, 34]. However, a precise estimation of heat transfer coefficient has not been mentioned in these experiments. Several theoretical models have been proposed that consider effect of precursory cooling [10, 14, 15, 16, 18] and an adiabatic condition in the dry region ahead of wet front [4, 6, 24]. These models [4, 6, 10, 16, 24] consider an approximate value of heat transfer coefficient (Bi) to correlate with the test data. The results of the present solution (Eq. 49) is characterized by various modeling parameters such as: zone of precursory cooling (α), magnitude of precursory cooling (N), Biot number (Bi), Peclet number (Pe) and dry wall temperature (θ_1). In this study, B_i, θ_1 and α are kept constant a priori with an adjustable value of N for comparison with the available test data taken from various sources and is shown in Figure 9. These test data cover a wide range of coolant flow rate, dry wall temperature, coolant type and different thicknesses of the test geometry. The result shows that the magnitude of precursory cooling (N) decreases with the increase in the coolant flow rate. It may be noted that with the increase in coolant flow rate, number of coolant droplets increases in the sputtering region, increasing the heat transfer rate from the dry region ahead of wet front. This yields a stronger precursory cooling and N decreases. Previous researchers have observed similar results in their analysis [10, 14, 18].

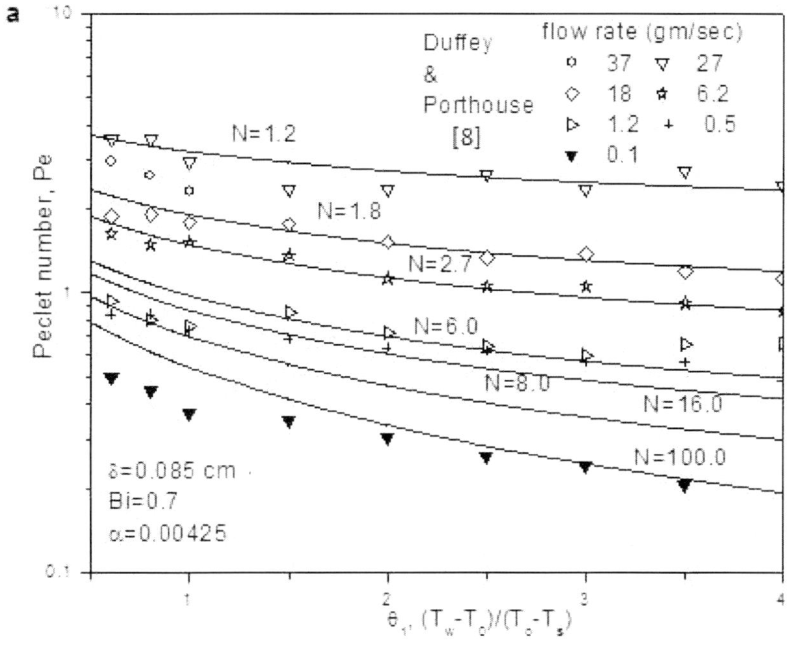

Figure 9. Validation of predicted wet front velocity with experimental results of Duffey and Porthouse [8] (wall thickness 0.085 cm).

In the present study, the effect of temperature dependent thermal conductivity in the rewetting model is analyzed by considering the variation of thermal conductivity of the hot solid to be a linear function of temperature. The solution is obtained by employing the variational integral method. The result obtained by the employing VIM analysis is compared with other available results. In the present study, present results are compared with analytical solution of Yamanouchi [24] and are presented in Table 1. It is seen that variational integral method predicts a lower value of Peclet number (Pe) compared to Yamanouchi [24] model for lower drywall temperature. The maximum error in the variation of Peclet number (Pe) between the present solution and Yamanouchi model [24] is found to 0.56% and 10.41% at $theta_1 = 1.0$ and $theta_1 = 7.0$, respectively. This shows that the effect of variable thermal conductivity on the rewetting rate increases at higher dry wall temperature.

The effect of temperature dependent thermal conductivity has been found to have minimal effect on the rewetting velocity. Following the success of these models the VIM technique has been extended to solve the multi-region rewetting problem with precursory cooling (Eq. 105, 106, 111, 112) and the obtained results are discussed here.

Table 1. Comparison between wet front velocity obtained by (a) Yamanouchi [24] and (b) present result for $\varepsilon_0 = 0.0012, Bi = 2.0$ [23]

	I	II	I & II
θ_1	Yamanouchi[a] [24]	VIM^b	Relative error 'E' (in %)
	Pe	Pe	
1	1	0.9943	0.5632
2	0.5773	0.5678	1.6413
3	0.4082	0.3957	3.0495
4	0.316	0.3019	4.523
5	0.2581	0.2421	6.1963
6	0.2182	0.2004	8.1488
7	0.1889	0.1692	10.4178
a: Yamanouchi [24]; One-dimensional model			
b: Result obtained by employing Variational Integral Method (VIM) [23]			

In the present study variational integral method proposed by Arpaci [3] is employed to analyze the rewetting behavior of a hot surface. The hot solid is divided into three distinct regions, namely, dry region ahead of quench front, sputtering region behind the wet front and a continuous liquid film region (Figure 5). A hybrid profile involving exponential and polynomial functions are used in the wet and dry region, while, in the sputtering region, three different temperature profiles involving different orders of polynomial functions have been tried. In all the cases, a closed form expression is obtained for the temperature field in different regimes, namely, wet, sputtering and dry region. Peclet number is found to depend on various modeling parameters, namely, drywall temperature, incipient boiling temperature, sputtering length, convective Biot number and boiling Biot number (Eq. 78, 79, 82,83, 86, 87).

A comparison among Peclet number obtained by various model for the special case of $Bi_C = Bi_B = Bi = 1$ and $\theta_2 = 0$ is shown in Table 2. For all the models, Peclet number increases with increasing Biot number. The results obtained by employing variational integral method using a parabolic temperature profile in the sputtering region (Model 1) is found to be in good agreement with the 2-D solution obtained by Bonakdar et al. [27]. However, present model underpredicts the result obtained by Sun et al. [5] and Sahu et al. [26] by employing Winer-Hopf technique and HBIM, respectively. The model that uses cubic polynomial guess function (model-3) predicts a lower Peclet number compared to the Peclet number obtained by model-1 and model-2.

It is to be noted that parameter θ_1 characterizes the difference between dry wall temperature and the quench front temperature. Dry wall temperature found to affect the velocity of wet front propagation. While, θ_2 represents the incipient boiling temperature that characterizes the length of sputtering region. Here, θ_2 ranges from zero to unity. For both the extreme values of θ_2, the sputtering length vanishes and the three-region model reduces to two-region rewetting model with a uniform heat transfer coefficient over the entire wet region. In a further refinement over the previous multi-region analysis (Figure 5), we have incorporated the effect of precursory cooling in the dry region ahead of wet front (Figure 6) and the obtained results are compared with available test data for a varied range of experimental data and test geometry. Numerous experimental investigations have been carried out to analyze the rewetting behavior of hot surfaces in different geometries such as: slab, rod, and tube [24, 30, 32]. Tests have been carried out either for a single rod [32] or with rod bundle [24, 30, 34] with varied range of coolant flow rate and dry wall temperature. It is observed that most of the theoretical models considered an approximate value of heat transfer coefficient (Biot number) to compare predicted models with the test data [4, 29]. Most of the experimental studies use water as a coolant under atmospheric condition. In such a case, θ_2 may be assumed to be smaller than 0.2 as reported by Sun et al. [5]. For lower values of $\theta_2 (\theta_2 < 0.2)$, the heat transfer in the continuous liquid region exerts little influence on the wet front velocity and the rewetting model can be considered as a two-region basic rewetting model. It is observed that by reducing the present model into a two-region model (neglecting sputtering zone) both the models (model-1 and model-2) predict same results. The results found and correlate the test data for higher flow rates. This may be due to the higher boiling Biot number used in the entire wet region.

The results obtained by employing model-2 (neglecting sputtering region) is compared with the experimental data taken from various sources with different wall thickness, test geometry and coolant flow rates and are shown in Figure 10(a-c). The experimental data are taken from water-stainless steel pair, considering water as the coolant. It is reported that the value of N and "a" are obtained with the knowledge of variation of heat flux in the dry region ahead of the wet front for a nitrogen-copper pair [31]. The value of N is considered as unity and the zone of precursory cooling was considered to be 0.1 for comparing the theoretical model with the test data [37]. However, the authors have varied the Biot number for correlating the test data at various coolant flow rates. The detailed information on the variation of heat flux in the dry region ahead of wet front for water-stainless steel pair is not reported in the literature. In the present prediction, the value of "a" is taken as $0.05cm^{-1}$ for comparing the experimental results. Earlier researchers have followed same procedure to compare the theoretical model with test data [10, 14].

Table 2. Comparison of present analysis with available analytical result [23]

$\theta_2 = 0, Bi_B = Bi_C = Bi = 1.0$							
	I	II	III	IV	V	VI	VII
	Eq.(78)	Eq.(82)	Eq.(86)	2-D	1-D	2-D	
θ_1	Pe	Pe	Pe	Pe	Pe	Pe	Pe
0.5	0.8956	0.0223	0.6957	1.7326	2.0018	1.2750	0.1711
1	0.5494	0.0114	0.4249	1.2253	1.4147	0.7326	0.1110
2	0.3172	0.0060	0.2447	0.8662	1.0001	0.3480	0.0733
3	0.2245	0.0041	0.1728	0.7072	0.8165	0.2337	0.0585
4	0.1741	0.0032	0.1338	0.6124	0.7072	0.1834	0.0501
I. Present solution (model-1)				V. Sun et al. [10]			
II. Present solution (model-2)				VI. Bonakdar and McAssey [27]			
III. Present solution (model-3)				VII. Casamirra et al. [28]			
IV. Sahu et al. [26]							

The experimental data considered in the present analysis covers a wide variation of coolant flow rate and wall thickness. The present prediction (model-2) provides an excellent agreement with the available test data of Duffey and Port-

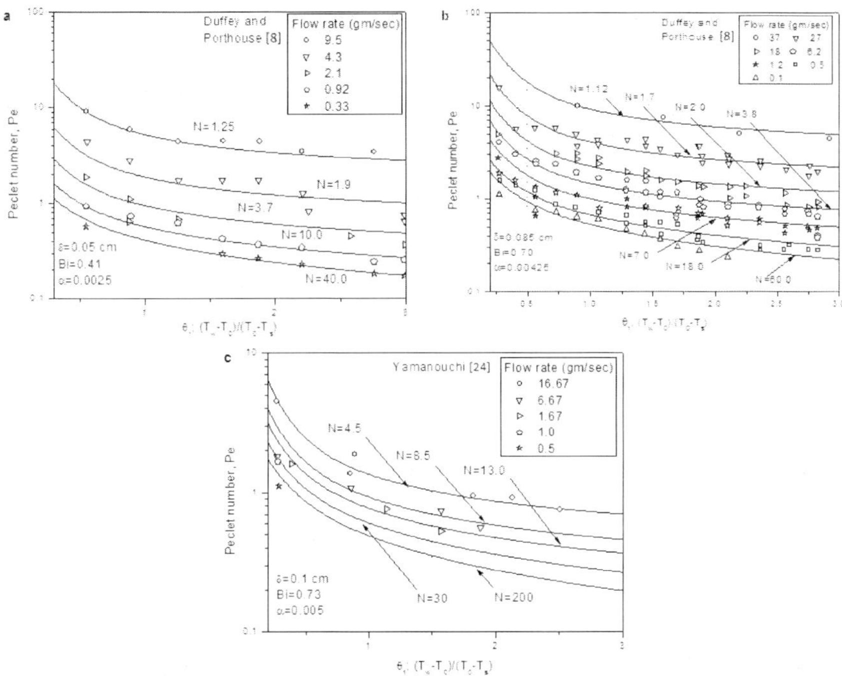

Figure 10. Comparison of wet front velocity obtained by VIM with experimental results of (a) Duffey and Porthouse [8] (wall thickness 0.05 cm) (b) Duffey and Porthouse [8] (wall thickness 0.085 cm) and (c) Yamanouchi [24] (wall thickness 0.1 cm) [25].

house [8] and Yamanouchi [24] for varied range of test geometry, coolant flow rate and dry wall temperature and is shown in Figure 10(a-c).

Based on the analysis, a correlation between magnitude of precursory cooling (N) and flow rate per unit perimeter (Ψ) has been proposed. In past, various researchers have proposed suitable correlation between coolant flow rate and N while comparing with the test data of Duffey and Porthouse [8] and Yamanouchi [24]. A summary of these correlations are presented in Table 3.

Table 3. Summary of correlations between flow rate per unit perimeter (Ψ) and magnitude of precursory cooling (N) [25]

Source	N as a function of flow rate per unit perimeter(Ψ))	Solution method	Model description
Sun et al. [10] (1-D)	$N=800\Psi^{1.4}$	Analytical	Exponential variation of heat transfer coefficient in dry region
Dua and Tien [14] (2-D Slab)	$N=(160\Psi)+1$	Winer-Hopf technique	Exponential variation of heat flux parameter in dry region
Sahu et al. [18] (2-D Slab, Cylinder)	$N=24.01\Psi^{0.701}$	HBIM	Exponential variation of heat flux parameter in dry region
Agrawal and Sahu [25] (1-D)	$N=18.60\Psi^{0.4947}$	VIM	Axially varying heat transfer coefficient in sputtering region and exponentially decaying heat transfer coefficient in dry region

Conclusion

Variational Integral Method has been employed to solve the heat conduction equation in host of rewetting problems. Initially, a one-dimensional rewetting model involving two regions, namely, wet region and dry region along the hot solid is considered for the analysis. Later on, the analysis is extended to include the effect of precursory cooling, property variation and axial variation in heat transfer coefficient in the basic rewetting model. In all the cases a closed form solutions have been obtained for temperature field and rewetting velocity of the hot solid. It has been observed that the effect of temperature dependent thermal conductivity has a minimal effect on the rewetting velocity. It may be noted that VIM has the ability to solve variety of rewetting problems without recourse to mathematical complexity and numerical computation. This technique is simple and can be adopted as a tool for solving conduction equation in host of rewetting problems. This study has shown that VIM has the potential to be used as alternative analytical technique for solving conduction problems.

References

[1] Biot M. A., New methods in heat flow analysis with application to flight structures, *J. Aero. Sci.* **24**, 857-873 (1957).

[2] Sparrow E. M., Siegel R., A variational method for fully developed laminar heat transfer in ducts, *J. Heat Transf. Trans. ASME* **81-C**, 157-166 (1959).

[3] Arpaci V. S., *Conduction Heat Transfer*. Reading, MA, USA: Addison-Wesley Publishing Company, 435-478, 1966.

[4] Sahu S. K., Das P. K., Bhattacharyya S., A comprehensive analysis of conduction-controlled rewetting by the Heat Balance Integral Method, *Int. J. Heat Mass Transf.* **49**, 4978-4986 (2006).

[5] Sun K. H., Dix G. E., Tien C. L., Cooling of a very hot vertical surface by falling liquid film, *ASME J. Heat Transf.* **96**, 126-131 (1974).

[6] Agrawal M. K., Sahu S. K., Analysis of conduction-controlled rewetting of a hot surface by variational method, *Heat Mass Transf.* **49**, 7, 963-971 (2013a).

[7] Rosenthal D., The theory of moving sources of heat and its application to metal treatments, *Trans. ASME* **68**, 11, 849-866 (1946).

[8] Duffey R. B., Porthouse D. T. C., The physics of rewetting in water reactor emergency core cooling, *Nucl. Eng. Des.* **25**, 379-394 (1973).

[9] Thompson T. S., An analysis of the wet-side heat transfer coefficient during rewetting of a hot dry patch, *Nucl. Eng. Des.* **22**, 212-224 (1972).

[10] Sun K. H., Dix G. E., Tien C. L., Effect of precursory cooling on falling-film rewetting, *ASME J. Heat Transf.* **97**, 360-365 (1975).

[11] Hsu C. H., Chieng C. H., Hua T., Two-dimensional analysis of conduction-controlled rewetting with internal heat generation, *in Proceedings of International Conference on Numerical Methods in Engineering*, Montreal, Canada (1983).

[12] Salcudean M., Bui T. M., Heat transfer during the rewetting of hot horizontal channels, *Nucl. Eng. Des.* **108** 323330 (1988).

[13] Simopoulos S. E., The effect of precursory cooling on the rewetting rate, *Atomkernenergie-Kerntechnik*. **49**, 37-42 (1986).

[14] Dua S. S., Tien C. L., Two dimensional analysis of conduction-controlled rewetting with precursory cooling. *ASME J Heat Transf.* **98**, 407-413 (1976).

[15] Olek S., The effect of precursory cooling on rewetting of slab, *Nucl. Eng. Des.* **108**, 323-330 (1988).

[16] Olek S., Wiener-hopf technique solution to a rewetting model with precursory cooling. *Nucl. Sci. Eng.* **105**, 271-277 (1990).

[17] Olek S., Rewetting of a solid cylinder with precursory cooling, *App. Scienti. Res.* **46**, 347-364 (1989).

[18] Sahu S. K., Das P. K., Bhattacharyya S., Rewetting analysis of hot vertical surfaces with precursory cooling by the heat balance integral method, *ASME J. Heat Transf.* **130** 2, 024504 (2008).

[19] Elias E., Yadigaroglu G., A general one-dimensional model for conduction controlled rewetting of a surface, *Nucl. Eng. Des.* **22** 42, 185-194 (1977).

[20] Agrawal M. K., Sahu S. K., Conduction-controlled rewetting analysis of a hot surface by variational method, *in 22nd National and 11th International ISHMT-ASME Heat and Mass Transfer Conference*, IIT Kharagpur, India (2013).

[21] Olek S., Zvirin Y., The effect of temperature dependent properties on the rewetting velocity, *Int. J. Multi. Flow* **11** 4, 577-581 (1985).

[22] Sahu S. K., Das P. K., Bhattacharyya S., How good is goodmans heat balance integral method for analyzing the rewetting of hot surfaces? *Therm. Sci.* **13** 2, 97-112 (2009).

[23] Agrawal M. K., Sahu S. K., Analysis of three-region conduction controlled rewetting of a hot surface by variational method, *Heat Trans. Eng.* **37**, 9 763-773 (2016).

[24] Yamanouchi A., Effect of core spray cooling in transient state after loss of coolant accident, *J. Nucl. Sci. Tech.* **5**, 547-558 (1968).

[25] Agrawal M. K., Sahu S. K., Analysis of multi-region conduction controlled rewetting with precursory cooling by variational method, *Applied Thermal Eng.* **73**, 265-274 (2014).

[26] Sahu S. K., Das P. K., Bhattacharyya S., A three-region conduction-controlled rewetting analysis by the heat balance integral method, *Int. J. Therm. Sci.* **48** 11, 2100-2107 (2009).

[27] Bonakdar H., McAssey Jr E. V., A method for determining rewetting velocity under generalized boiling conditions, *Nucl. Eng. Des.* **66** 1, 7-12 (1981).

[28] Casamirra M., Castiglia F., Giardina M., Lombardo C., Celeta G. P., Mariani A., Saraceno L., Rewetting of a hot verticle surface by liquid sprays, *Exp. Therm. Flu. Sci.* **29**, 885-891 (2005).

[29] Duffey R. B., Porthouse D. T. C., Experiment on the cooling of high temperature surfaces by water jets and drops, Report No. *RD/B/N2386*, Berkeley Nuclear laboratories, USA (1972).

[30] Sahu S. K., Das P. K., Bhattacharyya S., An experimental investigation on the quenching of a hot vertical heater by water injection at high flow rate, *Nucl. Eng. Des.* **240** 6, 1558-1568 (2010).

[31] Patil N. D., Das P. K., Bhattacharyya S., Sahu S. K., An experimental assessment of cooling of a 54-rod bundle by in-bundle injection, *Nucl. Eng. Des.* **250**, 500-511 (2012).

[32] Coney M. W. E., Calculations on the rewetting of hot surfaces, *Nucl. Eng. Des.* **31**, 246-259 (1974).

[33] Filipovic J., Incropera F. P., Viskanta R., Quenching phenomena associated with a water wall jet: I. Transient Hydrodynamic and Thermal Conditions, *Exp. Heat Transf.* **8**, 97-117 (1995a).

[34] Tien C. L., Yao L. S., Analysis of conduction-controlled rewetting of a vertical surface, *ASME J. Heat Transf.* **97**, 161-165 (1975).

[35] Anderson J. G. M., Hansen P., Two-dimensional heat conduction in rewetting phenomena, Report No. *NORHAVD-6*, Riso, Denmark, (1974).

[36] Chen W. J., Lee Y., Groeneveld D. C., Measurement of boiling curves during rewetting of a hot circular duct, *Int. J. Heat Mass Transf.* **22**, 973-976 (1978).

[37] Dua S. S., Tien C. L., An experimental investigation of falling-film rewetting, *Int. J. Heat Mass Transf.* **21**, 955-965 (1978).

[38] Sibamoto Y., Maruyama Y., Nakamura H., Measurement and analysis for rewetting velocity under post-BT conditions during anticipated operational occurrence of BWR, *ASME J. Eng. Gas Turb. Pow.* **132**, 102909-1 (2013a).

In: Heat Conduction
Editors: J. Hristov and R. Bennacer
ISBN: 978-1-53614-673-8

Chapter 3

THE HEAT RADIATION DIFFUSION EQUATION WITH MEMORY: CONSTITUTIVE APPROACH AND APPROXIMATE INTEGRAL-BALANCE SOLUTIONS

Jordan Hristov*
Department of Chemical Engineering,
University of Chemical Technology and Metallurgy,
Sofia, Bulgaria

Abstract

A constitutive approach to modelling of heat shock waves described by the diffusion approximation in radiation heat transfer in terms of time fractional derivatives has been developed. Approximate closed-form analytical solutions by a double integration method concerning a step change of the surface temperature and two classical problems with time-dependent boundary conditions have been developed.

Keywords: radiation heat transfer, heat conduction, memory, fractional derivatives

*Corresponding Author Email: jordan.hristov@mail.bg.

1. Introduction

Shock wave of heat propagation into solids caused by strong radiation is reasonably approximated by a nonlinear diffusion approximation widely applied in various areas of energy transfer such as astrophysics [1], high temperature plasma [2, 3], insulations [4], etc. Generally such a diffusion approximation relates the heat flux at any point to its local gradient, which is a classical approach for the so-called simple materials [5], and known as the Fourier law. Under the conditions of high heat flux [6] the energy transfer by radiation and absence of fluid motion is modelled by one-dimensional relationships, namely

$$\rho \frac{\partial e}{\partial t} = \frac{\partial \Phi}{\partial x}, \quad \Phi = -\frac{4}{3K\rho} \frac{\partial \sigma T^4}{\partial x} \tag{1}$$

The relation of the flux to the as a gradient of the 4th power of the local temperature is the Rosseland approximation [1, 5, 6] applicable only to thick media and absence of fluid flow [5, 6, 7, 8, 9]. The solution heat radiation diffusion equation (1) was a challenging problem solved in different ways, among them: self-similar solution [10, 11], traveling wave approach [12]. Related problems with non-linear heat conduction have developed by the integral balance approach [13, 14] with well defined convex shapes of the temperature profiles moving as shock waves with finite speeds. Other classical approach is to express the solution as series of a dimensionless group (preliminary suggested as anzatz) in a structure of a series parabolic profiles.

2. The Nonlinear Heat Radiation Diffusion Equation: Classical Formulation and Memory Fffects

2.1. The Heat Radiation Diffusion without Inertia (without Flux Relaxation)

The internal energy and the Rosseland mean opacity depend on the temperature and density as power-laws [6]

$$e = f \frac{T^{\beta}}{\rho^{\mu_1}}, \quad \frac{1}{K} = g \frac{T^{\alpha}}{\rho^{\lambda}} \tag{2}$$

For gold, for instance, we have $g = 1/7200 g/cm^2$ [12]. Moreover, $\alpha > 0$ and $\beta > 0$ are constants and approximately [6, 12],$\alpha_1 = 1.5$ and $\beta = 1.6$. With the power-law relationships (2) the thermal balance equations (1) results in

$$\frac{\partial T^{\beta}}{\partial t} = a\frac{\partial^2 T^{4+\alpha}}{\partial^2 x}, \quad a = \frac{16\epsilon}{\beta}\frac{g\sigma}{3f\rho^{2-\mu_1+\lambda}}, \quad \epsilon = \frac{\beta}{4+\alpha} \tag{3}$$

In equation (3) the coefficient of thermal diffusion a has a dimension $\left[m^2/sK^{(4+\alpha-\beta)}\right]$ but for the special case when $\alpha = 0$ and $\beta = 1$ its dimensions reduces to $a \equiv \left[m^2/sK^3\right]$. Moreover, this model suggests that the thermal diffusivity a is temperature independent since the principle assumptions are constant material density ρ and heat capacity C_p, and constant heat conductivity $k \neq k(T)$.

By changing the variables $U = T^{4+\alpha-\beta}$ we may present (3) as an equation including both wave and diffusion effects [6]

$$\frac{\partial U}{\partial t} = a\left[\frac{1}{1-\epsilon}\left(\frac{\partial U}{\partial x}\right)^2 + \frac{U}{\epsilon}\frac{\partial^2 U}{\partial x^2}\right] \tag{4}$$

In all cases it should be considered $\epsilon \neq 1$, since $\epsilon = 1$ corresponds to the linear diffusion model. Equation (4) was solved by a series approximation [6] with defined front separating the disturbed and virgin area ($T = 0$ for $x \geq x_F$, that is $U(0,t) = 0$) of the medium

$$U(z,t) = \sum_{i=1}^{\infty} c_i(t) z^i, \quad z = x_F - x \tag{5}$$

The formulation of (3) may be represented as

$$\frac{\partial \theta}{\partial t} = a\frac{\partial^2 \theta}{\partial x^2}, \quad \theta = T^{\beta}, \quad \gamma = \frac{4+\alpha}{\beta} \tag{6}$$

This is a degenerate parabolic equation with a common form

$$\frac{\partial \theta}{\partial t} = \frac{\partial}{\partial x}\left[a_{\gamma}\theta^{\gamma-1}\frac{\partial \theta}{\partial x}\right] \tag{7}$$

The meaning of the presentation (7) is that the heat flux depends non-linearly from the gradient, namely

$$q = -k(\theta)\frac{\partial \theta}{\partial x} \tag{8}$$

through the power-law relationship of the thermal diffusivity $a(\theta) = a_\gamma \theta^{\gamma-1}$ where $a_\gamma = a/\gamma$. Then the continuity equation (the energy balance) (9) yields the diffusion equation (7)

$$\frac{\partial \theta}{\partial t} = -\frac{\partial q}{\partial x} \tag{9}$$

In terms of the original variable T we have (10) which consequently leads to (3)

$$q = -k(T)\frac{\partial T^\beta}{\partial x}, \quad a_0 = \frac{k_0}{\rho C_p}, \quad k(T) = k_0 \frac{\beta}{4+\alpha} T^{(4+\alpha)} \tag{10}$$

The relationship (10) is based on the assumption that both the material density ρ and the heat capacity C_p are temperature independent while the heat conductivity $k(T)$ is temperature-dependent that finally results in a temperature dependent thermal diffusivity as $a = a_0 T^{(4+\alpha)}$. This is the Fourier law where an instantaneous heat flux with infinite speed is defined. Despite the fact that solutions of the degenerate diffusion equations such as (7) move with well defined fronts [12, 13, 14, 16] a fact coming from the equivalence of (7) to the wave equation (4) in the heat flux propagation into the medium there is no damping effect (time relaxation) which is a direct consequence the application of the Fourier law.

2.2. Memory Formalism and Flux Damping (Relaxation)

Actually, the flux propagation into the medium should be related to its history through the integral

$$q = -\int_0^t R_D(t-s)a(T)\frac{\partial T^\beta(x,s)}{\partial x}ds, \quad R_D(t-s) = \delta_D(t) \tag{11}$$

In the formulation (11) the *influential function* $R_D(t-s)$, termed here after also as a *memory kernel* relates the present sate of the temperature gradient to its history. In this context, in the original heat radiation diffusion model (3) and all its equivalent representations the memory kernel is equals the direct Delta function so that $\int_0^t \delta_D(s)ds = 1$. This actually means that no damping effect on the heat flux propagation into the medium exists in accordance with this model formulation. The memory function should be positive and attain zero for long times. Two principle memory functions are commonly used [5, 17]:

a)$R_D(s) = s^r(s+1)^{-\mu}$ is memory function of order r, if $r < \mu$. For $r = 0$ we have a singular memory kernel which leads to applications of fractional integral and derivatives. This kernel is *scale free* or *scale invariant function*.

b) The exponential function $e^{\beta s}$ is memory kernel which is not scale invariant. This memory function is directly related to the fractional operators of Caputo-Fabrizio [18, 19].

The models of heat radiation diffusion with memory developed in this chapter consider only singular memory kernels and application of classical fractional operators of Riemann-Liouville and Caputo [20]. The basic formulations are presented briefly next.

2.3. The Heat Radiation Diffusion with Inertia: Model Formulation

From the previous point regarding the retardation (memory effect) in the heat flux propagation we consider power-law function (12) as a memory kernel

$$R(t) = \frac{t^{\mu-1}}{\Gamma(\mu)} \tag{12}$$

This memory kernel corresponds the well-known fractional operators with singular kernels [20] such as the Riemann-Liouville fractional integral (13) and consequently to time fractional derivatives of Riemann-Liouville (14) and Caputo (15) sense

$${}^{RL}I_t^{\mu} = D^{1-\mu}[f(t)] = \frac{1}{\Gamma(\mu)} \int_0^t \frac{f(x,s)}{(t-s)^{1-\mu}} ds \tag{13}$$

$${}^{RL}D_t^{\mu} f(x,t) = \frac{1}{\Gamma(1-\mu)} \frac{d}{dt} \int_0^t \frac{f(x,s)}{(t-s)^{\mu-1}} ds \tag{14}$$

$${}^{C}D_t^{\mu} f(x,t) = \frac{1}{\Gamma(1-\mu)} \int_0^t \frac{1}{(t-s)^{\mu-1}} \frac{df(x,s)}{dt} ds \tag{15}$$

The choice of the power-law memory kernel is due *the rapid* (but not with infinite sped) *heat flux propagation*, that is *short time-memory effects* and to the fact that this approach leads to the non-linear time-fractional diffusion (heat conduction equation). Hence, the flux q is related to the temperature gradient dT^{β}/dx through a the power-law memory kernel in the form of the Riemann–Liouville fractional Integral (13), namely

$$q = -{}^{RL}I_t^{\mu}\left[a(t)\frac{\partial T^{\beta}}{\partial x}\right] = D_t^{1-\mu}\left[\frac{\partial T^{\beta}}{\partial x}\right] = \frac{1}{\Gamma(\mu)} \int_0^t \frac{1}{(t-s)^{1-\mu}} \frac{\partial T^{\beta}(x,s)}{\partial t} ds \tag{16}$$

Then applying the continuity equation

$$\frac{\partial T^{\beta}}{\partial t} = -\frac{\partial q}{\partial x} \tag{17}$$

we get the non-linear heat diffusion equation with a memory i.e., non-linear time-fractional heat diffusion equation

$$\frac{\partial^{\mu}}{\partial t^{\mu}} T^{\beta} = \frac{\partial}{\partial x}\left[\left(a_0 \frac{\beta}{4+\alpha} T^{(4+\alpha)}\right)\frac{\partial T^{\beta}}{\partial x}\right], \quad a_0 = \frac{k_0}{\rho C_p}, \quad 0 < \mu < 1 \tag{18}$$

With the substitution $\theta = T^{\beta}$ that means $T = \theta^{\frac{1}{\beta}}$ and $T^{(4+\alpha)} = \theta^{\frac{4+\alpha}{\beta}}$ we get

$$\frac{\partial^{\mu}\theta}{\partial t^{\mu}} = \frac{\partial}{\partial x}\left[\left(a_{\gamma}\theta^{\gamma}\right)\frac{\partial \theta}{\partial x}\right], \quad 0 < \mu < 1 \tag{19}$$

where

$$a_{\gamma} = a_0 \frac{\beta}{4+\alpha}, \quad a_{\gamma} = \frac{a_0}{\gamma}, \quad \gamma = \frac{\beta}{4+\alpha} \tag{20}$$

In (18) and (19) the symbol $\frac{\partial^{\mu}}{\partial t^{\mu}}(\bullet)$ denotes either Riemann-Liouville or Caputo time-fractional derivative [20].

The time-fractional subdiffusion equation (19) was solved recently by the integral-balance approach [21] and this solution methodology will be demonstrated in this chapter.

3. Solution Approach: A Necessary Background about the Techniques Used

3.1. Integral Balance Method

The integral-balance approach uses the concept of a final penetration depth $\delta(t)$ of the diffusant (heat or mass) which eliminate *ad hoc* the infinite flux speed of the classical diffusion equation. With this concept, in a semi-infinite medium, the conditions at infinity ($T(\infty) = 0$ and $\partial T(\infty, t)/\partial x = 0$) can be replaced by condition at the impermeable front, namely

$$T(\delta) = 0, \quad and \quad \frac{\partial T}{\partial x}(\delta) = 0 \tag{21}$$

The new conditions from a sharp front $\delta(t)$ with unknown position that should be determined as part of the solution. The classical integral-balance method applied to the heat conduction equation with a thermal- independent diffusivity is (i.e.$a = a_0 = const.$) [13, 14, 16].

$$\frac{\partial T}{\partial t} = a_0 \frac{\partial^2 T}{\partial x^2} \tag{22}$$

Consider integration over the finite penetration depth δ of the diffusion (thermal) layer. Applying the Leibniz rule we get (23)

$$\frac{d}{dt} \int_0^{\delta} T(x,t)dx = -a_0 \frac{\partial T}{\partial x}(0,t) \tag{23}$$

The integral relation (23) is a simple energy balance, since the left side of (23) is the time variation of the energy accumulated in the thermal layer, which is controlled by the flux at the boundary (the right side of (23)). This is the simplest version of the integral method known as *Heat-balance Integral Method* (HBIM) [22]. The core of the approximate solution is the replacement of T by an assumed profile T_a expressed as a function of the relative space co-ordinate x/δ). The result of this operation is an ordinary differential equation about $\delta(t)$ [13, 14, 22, 23]. The principle disadvantage of HBIM is that $\partial T_a(0,t)/\partial x$ is dependent on the type of the assumed temperature profile [13, 14, 22, 23].

However, if a double integration is carried out in accordance with the *Double Integration Method* (DIM) [13, 14], then the right-side part of the integral relation depends only on the boundary condition $T(0,t)$ and is unaffected by the choice of the assumed temperature profile. The integral relation of DIM is [14]

$$\int_0^{\delta} \left(\int_x^{\delta} \frac{\partial T}{\partial t} dx \right) dx = a_0 T(0,t) \tag{24}$$

This integral relation allows to work with either integer-order time-derivatives [13, 14] or with time-fractional derivatives [21, 24](as in the present case) where the Leibniz rule is inapplicable.

Now, let us turn on the problem related to the heat radiation diffusion equation at issue. When the thermal diffusivity is power-law dependent on the temperature $a = a_p T^m$ $(m > 0)$, then the integral relation corresponding to DIM [13, 14] is

$$\int_0^\delta \left(\int_x^\delta \frac{\partial T}{\partial t} dx \right) dx = \frac{a_p}{m+1} \left[T(0,t) \right]^{m+1} \tag{25}$$

3.2. Assumed Profile

The approximate integral balance solutions presented in this chapter use an assumed parabolic profile with unspecified exponent [13, 14, 16, 23, 24], generally expressed as

$$u_a = u_s \left(1 - \frac{x}{\delta}\right)^n \tag{26}$$

where the variable $u_a(x,t)$ could be either $T(x,t)$ or $\theta(x,t)$, depending on the context the problem solved as it demonstrated in the sequel. This profile can be successfully applied to linear and non-linear diffusion problems with integer-order (local) [13, 14, 16] and fractional (non-local) time derivatives [21, 24]. This profile satisfies all boundary conditions imposed at the boundary of the diffusion layer for any positive value of the exponent n [23]. Moreover it allows to generate either concave (as in the classical diffusion model) or convex (like shock waves) solution profiles as it will demonstrated in this chapter. For more details about the application of the assumed profile (26) in the integral-balance solutions we refer to [12, 13, 14, 16, 21, 23, 24].

3.3. Transformation of the Diffusion Term and the Governing Equation to a Degenerate Equation

Following the above transformations of heat diffusion and for the sake of clarity of the further solutions we can present (3) as well as its equivalent form as

$$\frac{\partial \theta}{\partial t} = a \frac{\partial^2 \theta^w}{\partial x^2}, \quad w = \frac{4+\alpha}{\beta} \tag{27}$$

From (27) it is obvious that defined in (3) as well as $\epsilon = \gamma$. From the values of α and β mentioned at the beginning we have that $w > 1$ which allows (26) to expresses as [13]

$$\frac{\partial \theta}{\partial t} = \frac{\partial}{\partial x} \left[a_w \theta^m \frac{\partial \theta}{\partial x} \right], \quad \theta^m \frac{\partial \theta}{\partial x} = \frac{1}{m+1} \frac{\partial \theta^{m+1}}{\partial x} \tag{28}$$

where $a_w = w \cdot a$ and $m = w - 1 > 0$. Here we use the notation already used in [13, 21]. The transformed equation (28) is a slow diffusion model [25, 26] and allows easily the Double-Integral Method to be applied. Equation (28 degenerates at $\theta = 0$, that is at the front of the moving solution δ because for $x = \delta$ we have $\theta = 0$ and consequently $a(\theta) = 0$. A change of variables $\varphi = \theta^m$ and $\tau = t/m$ allows (28), to be transformed as

$$\frac{\partial \varphi}{\partial \tau} = a_w \left[\left(\frac{\partial \varphi}{\partial x} \right)^2 + m\varphi \frac{\partial^2 \varphi}{\partial x^2} \right] \tag{29}$$

This is equivalent to (4) and clearly demonstrates a simultaneous action of a wave-like (first term in the right-hand side) and a diffusion transport (second term) mechanisms. This transformed models was successfully solved by HBIM in [16].

4. DIM Solutions: Constant Density Cases

Prior to the development of the solution of the models with memory we like to draw a picture of the consequent part of this chapter. In order to demonstrate the new properties we will solve approximately by the double integral method three cases available in the literature, that is:

1) Transient problem with a step change in the boundary condition [27]

2) Marshak's problem with exponential growing temperature as a boundary condition [10, 27]

3) Garnier's problem with a power-law (in time) rising boundary temperature [9, 27].

The integer order versions of these three classical cases where solved by DIM recently [27] and we will used the solutions as reference ones comparing them to the new results presented in this chapter.

4.1. Step Change of Temperature at the Boundary

This problem clearly demonstrates the solution technique of DIM. With a Dirichlet boundary condition where with $T_s = const.$ we have $\theta_s = \theta(0, t) = T_s^\beta = const.$ Applying the DIM integral relation to (19) which is a fractional analog of (25) we get

$$\int_0^\delta \int_x^\delta \frac{\partial^\mu}{\partial t^\mu} \theta(x, t) dx dx = a_\gamma \frac{\theta^\gamma(0, t)}{\gamma + 1} \tag{30}$$

With the assumed parabolic profile (26) $\theta_a = \theta_s (1 - x/\delta)^n$ we obtain the following ordinary fractional differential equation about $\delta(t)$ [21, 28]

$$\frac{\partial^{\mu}\delta(t)}{\partial t^{\mu}} = a_{\gamma}\frac{\theta^{\gamma}(0,t)}{\gamma+1} \tag{31}$$

With either RiemannLiouville or Caputo derivative the solution of (31) with the initial condition is [21]

$$\delta^{\mu}(t) = \sqrt{a_{\gamma}t^{\mu}}\sqrt{\frac{(n+1)(n+2)}{\Gamma(1+\mu)(\gamma+1)}} \tag{32}$$

For $\mu = 1$ we get the integer-order solution [13, 16]

$$\delta_{a_{\gamma}(DIM)} = \sqrt{a_{\gamma}t^{\mu}}\sqrt{\frac{(n+1)(n+2)}{\gamma+1}} \tag{33}$$

Hence, the approximate solution is

$$\Theta^{\mu} = \frac{\theta_a}{\theta_s} = \left(1 - \frac{x}{\sqrt{a_{\gamma}t^{\mu}}\sqrt{\frac{(n+1)(n+2)}{\Gamma(1+\mu)(\gamma+1)}}}\right)^{n} \tag{34}$$

thus defining the non-Boltzmann similarity variable $\eta_{\mu} = x/\sqrt{a_{\gamma}t^{\mu}}$. With the inverse change of variables $T = \theta^{1/\beta}$ and $T_s = \theta_s^{1/\beta}$ we get

$$T_{norm} = \frac{T}{T_s} = \left(1 - \frac{x}{\sqrt{a_{\gamma}t^{\mu}}\sqrt{\frac{(n+1)(n+2)}{\Gamma(1+\mu)(\gamma+1)}}}\right)^{\frac{n}{\beta}} \tag{35}$$

The parabolic profile (35) depends on dimensionless ratio x/δ thus defining the similarity variable $\eta_{\mu} = x/\sqrt{a_{\mu 0}t^{\mu}}$ which for $\mu = 1$ reduces to the classical Boltzmann similarity variable $\eta_{\mu=1} = x/\sqrt{a_0 t}$. On the other hand, the profile (35) defines the *effective similarity variable* $\eta_{eff}(\mu,\gamma) = x/\sqrt{a_{eff}\mu\gamma t^{\mu}}$ with effective thermal diffusivity $a_{eff(\mu,\gamma)} = a_0/\left[(y+1)\Gamma(1+\mu)\right]$. Hence, the effective diffusivity can be presented as a product of a_0 and a retardation factor $p_{\mu,\gamma} = 1/\left[(y+1)\Gamma(1+\mu)\right]$ and $0 < p_{\mu,\gamma} < 1$. The effect of the parameter γ is stronger because $\Gamma(1+\mu)$ for $0 < \mu < 1$ has an order of magnitude of unity. With the effective thermal diffusivity the approximate solution is

$$\Theta_a^{\mu}(x,t) = \left(1 - \frac{x}{\sqrt{a_{eff(\mu,\gamma}t^{\mu}}\sqrt{(n+1)(n+2)}}\right)^n = \left(1 - \frac{\eta_{eff(\mu,\gamma)}}{\sqrt{(n+1)(n+2)}}\right)^n \tag{36}$$

The second version of (36) has the same form as the classical integer-order diffusion solution [29, 30] when $\mu = 1$.

As it was established in [13], for $\mu = 1$ the exponent of the solution (34) follows the rule $n = 1/\gamma$. Thus in this specific case $n_{(\mu=1)} = 1/\gamma\beta$ and with $\gamma = w - 1 = 4 - 1 = 3$, and $\beta = 1.6$ we have $1/\gamma\beta \approx 0.208$ [27]. In this case $\eta_{(\mu=1)} = x/\sqrt{a_{\gamma}t}$ defines the position of the front of the thermal wave.

However, when $0 < \mu < 1$ it is not possible to determine directly the exponent of the approximate profile but by help of the result developed in [21] we may define by interpolating the results as a function of μ for a given value of γ.

The residual function $R(x,t)$ of (28) when the approximate solution θ_a is used can be defined as

$$R_1(x,t) = \frac{\partial^{\mu}\theta_a}{\partial t^{\mu}} - \frac{\partial}{\partial x}\left(a_{\gamma}\theta_a^{\gamma}\frac{\partial \theta_a}{\partial x}\right) \tag{37}$$

The residual function should attain a minimum for a certain value of the exponent n since θ_a does not match the exact solution. The residual function can be easily evaluated by transformation of the moving boundary domain $0 \leq x \leq \delta(t)$ into one $0 \leq \xi \leq 1$ with fixed boundaries. By introduction of the Zener coordinate $\xi = x/\delta$ [31]. Now, the approximate profile can be presented as

$$\frac{\partial^{\mu}\theta_a}{\partial t^{\mu}} \longrightarrow \frac{\partial^{\mu}V(\xi,t)}{\partial t^{\mu}} = \frac{\partial^{\mu}}{\partial t^{\mu}}(1-\xi)^n \tag{38}$$

The approximation of the approximate solution (profile) $V(\xi) = (1-\xi)^n$ as a converging series $V_a(\xi) \approx \sum_{k=0}^{N} b_k\xi^k, 0 \leq \xi \leq 1$ [28] allows to carry out the time-fractional differentiation and to obtain approximation of the fractional derivative from the approximate profile. This is an alternating series which converges rapidly that easily can be proved by the Leibniz test. For example (up to 4 terms, i.e., for $N = 4$, we have

$$V_a(\xi) \approx 1 - n\xi + \frac{n(n-1)}{2}\xi^2 - \frac{n(n-1)(n-2)}{6}\xi^3 + \frac{n(n-1)(n-2)(n-3)}{24}\xi^4 + O(\xi^5) \tag{39}$$

Taking into account that $(\xi = \left(x/\sqrt{a_\gamma}\right) t^{\frac{\mu}{2}})/F_{n,\gamma}$. The denominator $F_{n,\gamma}$ is the numerical factor in the expression of $\delta(t)$. This allows the series (39) to be expressed as (we for $N = 3$, for the sake of simplicity)

$$V_a(\xi) \approx 1 - n\frac{x}{\xi}t^{-\frac{\mu}{2}} + \frac{1}{2}n(n-1)\xi^2 t^{-\mu} - \frac{1}{6}n(n-1)(n-2)\xi^3 t^{-\frac{3}{2}\mu} \quad (40)$$

Since all terms are power-law functions of time, we may apply easily to each of them the rule of fractional differentiation that leads to

$$_{RL}D_t^{\mu}V(\xi) \approx \sum_{k=0}^{N} c_k t^{k-\mu}, \quad c_k = b_k\left(\Gamma(1+k/\Gamma(1+k-\mu)\right) \quad (41)$$

Moreover, rearranging the terms in (41) so that each of them to contain a power of ξ we may express each term as a product of a function dependent on ξ and $t^{-\mu}$, that is [28]

$$_{RL}D_t^{\mu}V(\xi) \approx_{RL} \Phi(n,\xi,\mu)t^{-\mu} \quad (42)$$

Therefore, we may present the approximate residual function as (with the Riemann-Liouville derivative)

$$^{RL}R_1(\xi,t) = \left(_{RL}D_t^{\mu}\sum_{k=0}^{N} b_k\xi^k - a_{\mu 0}\left(\frac{m}{2}\right)V^{m-1}\left(\frac{\partial V}{\partial x}\right)^2\right) \quad (43)$$

Alternatively, with the Caputo derivative of the series expansion $V_a(\xi) \approx \sum_{k=0}^{N} b_k\xi^k$ we get $({}_{RL}\Phi(n,\xi,\mu)t^{-\mu} -_C \Phi(n,\xi,\mu)t^{-\mu}) = 1/\Gamma(1-\mu)$ [28] which is independent of n. The residual function has the same construction as that presented by (43) .

With the approximate residual function, at the boundary $x = 0$, we get

$$R_{RL}(0,t) = \left(\Phi(n,0,\mu)t^{\mu} - a_{\mu 0}\left(\frac{\gamma}{\delta^2}\right)n^2\right) = \frac{\bar{R}_{RL}(0,t)}{t^{\mu}} \quad (44)$$

$$^{RL}\overline{R}_1(0,t) = \left(\frac{1}{\Gamma(1-\mu)} - \gamma(\gamma+1)\Gamma(1+\mu)\frac{n^2}{(n+1)(n+2)}\right) \quad (45)$$

The solution of $\bar{R}_{RL}(0,t) = 0$, for various μ and γ provides values of n shown in Table 1. The value n decays (see Figure 1 as illustration) with increase in both μ and γ. In general we have $n < 1$ and the presented results are for large γ (we present only cases for $\gamma \geq 3$ due the conditions imposed by the heat radiation equitation. It is worthnoting that with $n < 1$ the assumed profile (26) generates convex plots as solutions [13, 16], and concave profiles for $n > 1$.

Table 1. Optimal exponents of the approximate profile at $x = 0$

$\gamma \downarrow, \mu \rightarrow$	0.1	0.2	0.3	0.4	0.5	0.6	0.7	0.8	0.9	0.99
3	0.577	0.557	0.524	0.481	0.429	0.369	0.303	0.231	0.150	0.042
4	0.408	0.395	0.374	0.345	0.310	0.269	0.223	0.172	0.113	0.032
5	0.316	0.306	0.290	0.269	0.243	0.212	0.177	0.137	0.091	0.026

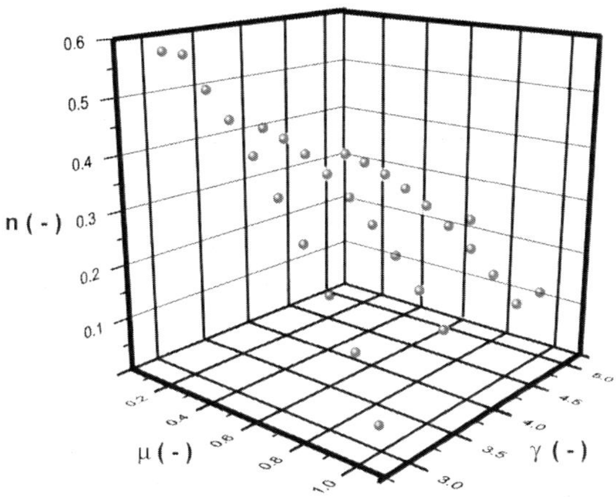

Figure 1. The exponent n defined at $x = 0$ as a function of μ and γ.

Hence, with $\gamma = 3$ we have a decaying exponent as the fractional order approaches unity. The behaviour is shown in Figure 1 for three integer values of γ and consequently the ratio n/β can be determined straightforwardly (see the data in Table 1).

The boundary conditions (21) ($\theta_a(\delta, t) = 0$ and $\partial\theta_a(, t)/\partial x = 0$) for $x \longrightarrow$

1 that means $\Phi(n,1,) \longrightarrow 1$ require [21, 24, 28]

$$\lim_{x \longrightarrow \delta} R(\delta,t) = 1 - D_{\mu 0}\left(\frac{\gamma}{\delta^2}\right) V^{\gamma-1}\left(\frac{\partial V}{\partial \xi}\right)^2 \equiv \frac{\gamma}{\delta^2}(1-\xi)^{n(\gamma-1)} n^2 (1-\xi)^{2(n-1)} \tag{46}$$

It follows from (46) that it could be satisfied at $x = \delta$ by the approximate profile if the exponent n is positive and satisfies the inequality [21, 24, 28]

$$n(\gamma-1) + 2(n-1) > 0 \Longrightarrow n > \frac{2}{\gamma+1} \tag{47}$$

The condition (46) and the estimations for n when $\xi \longrightarrow 1$ is practically the same as that observed when $\xi \longrightarrow 0$, that is n decreases with increase in γ. The effect of μ on n is almost the same but the effect of γ on the exponent is stronger.

The error measure defined as $E(n,\mu,m,t) = \int_0^\delta [R(x,t)]^2\, dx$ is, in fact, minimization of the residual function over the length of the thermal layer [32]. In ξ-space this means

$$E(n,\mu,\gamma,t) = \int_0^1 \left[\Phi(n,\xi,\mu)t^{-\mu} d\xi - a_\gamma \left(\frac{\gamma}{\delta^2}\right) n^2\right]^2 d\xi = \left[\frac{e(n,\gamma,\mu)}{\delta^2}\right] \tag{48}$$

In accordance the last form of (48) we have that E decreases in time because the denominators is $\delta^2 \equiv t^\mu$. The solution of (48), in fact, the minimization with respect to n provides optimal values of the exponent n_{opt} (see Table 2).

Table 2. Optimal exponents of the approximate profile

$\gamma \downarrow, \mu \rightarrow$	0.1	0.2	0.3	0.4	0.5	0.6	0.7	0.8	0.9	0.99
3	0.124	0.118	0.110	0.103	0.093	0.081	0.066	0.047	0.025	0.021
4	0.104	0.096	0.090	0.083	0.075	0.065	0.053	0.038	0.020	0.018
5	0.085	0.0812	0.076	0.070	0.063	0.055	0.044	0.032	0.017	0.016

The heat wave profiles expressed as functions of the similarity variable η_μ demonstrate shock wave behaviour. Precisely, with increase in the value of γ the effect of the nonlinearity in the thermal diffusivity dominates and controls the entire temperature distribution in the depth of the medium and is manifested by common convex shapes over the entire range $0 < \mu < 1$ (see Figure 2) . In general, with increase in γ the *thermal diffusion layer becomes shorter* and *the profiles become steeper* with almost vertical crossings with the abscissa. The latter is a well-known fact from the integer-order (without memory) degenerate diffusion equations [13, 14, 16, 25, 26].

4.2. Temperature-Independent Properties with Time-Dependent Boundary Condition (Marshak's Problem)

4.2.1. Marshak's Approach

Marshak [10] considered just a material heating and a model in the original notations (resembling (27))

$$\frac{\partial T}{\partial t} = \frac{D_M}{(p+4)} \frac{\partial T^{4+p}}{\partial x^2}, \quad D_M = \frac{4cl_0}{3} \tag{49}$$

The pseudo-diffusion coefficient D_M has a dimension $m^3/sK^{(3+p)}$ as it was demonstrated in eq. (3). Marshak [10] considered time-dependent boundary condition: $T_s = J_0 exp(2\alpha_a t)$ at $x = 0$ where J_0 is the initial surface temperature at $t = 0$ and α_s is a time constant.

Now, let consider the version of (49) with memory

$$\frac{\partial^\mu T}{\partial t^\mu} = \frac{D_M}{(p+4)} \frac{\partial T^{4+p}}{\partial x^2}, \quad D_M = \frac{4cl_0}{3} \tag{50}$$

Now the thermal diffusivity D_M has a dimension $m^3/s^\mu K^{(3+p)}$.

The application of DIM integral relation (25) to (50) with $T_a = T_s (1 - x/\delta)^n$, with $T_s = T_s(t)$ yields

$$D_t^\mu \left[J_0 exp(2\alpha_s t)\delta^2\right] = D_M \left[J_0 exp(2\alpha_s t\right]^{4+p}, \quad N(n,p) = \frac{(n+1)(n+2)}{(p+4)} \tag{51}$$

For $\mu = 1$ we have

$$\frac{d\left[J_0 exp(2\alpha_s t)\delta^2\right]}{dt} = D_M N(n,p) \left[J_0 exp(2\alpha_s t)\right] \tag{52}$$

and with the initial condition the result is [27]

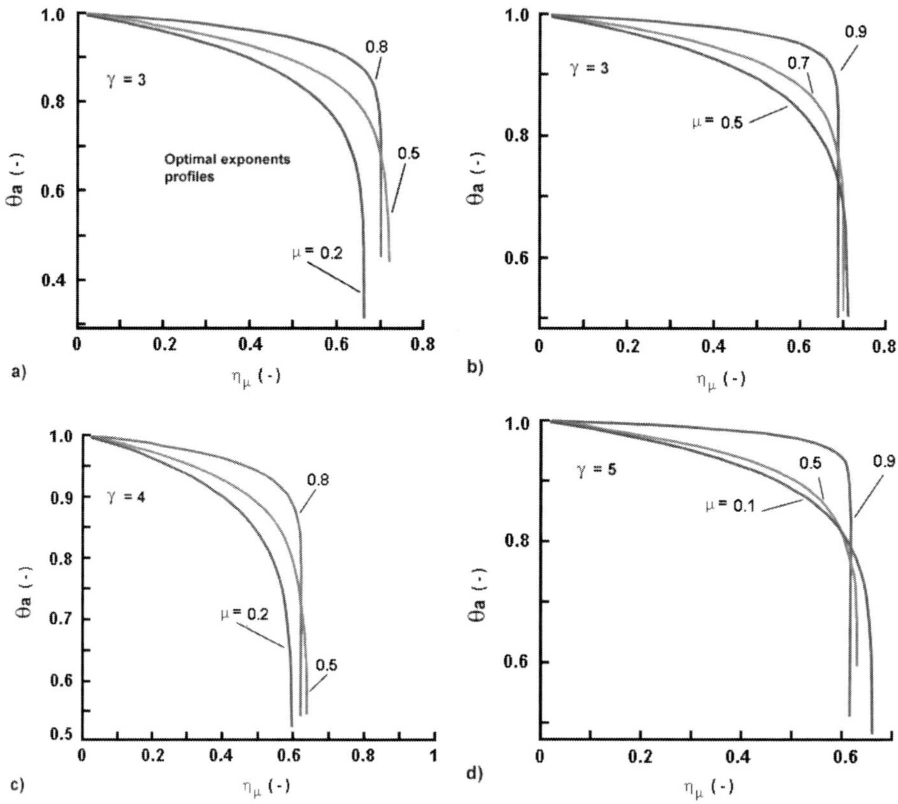

Figure 2. Approximate heat wave profiles developed by DIM and expressed as functions of the similarity variable η_μ. Strong effects of the nonlinearity γ of the thermal diffusivity diffusion coefficient ($\gamma \geq 3$) and suppressed memory (subdiffusion) effects . a) Approximate profiles of the heat wave for $\gamma = 3$ and $\mu = 0.2$, $\mu = 0.5$, $\mu = 0.8$. b) Approximate profiles of the heat wave for $\gamma = 3$ and $\mu > 0.5$ (weak memory effects). The fronts of the profiles are steeper that those in Figure 2a. c) Approximate profiles of the heat wave for $\gamma = 4$. The strong effect of γ suppressing the memory effects over the entire range $0 < \mu < 1$ and resulting in profiles with steep fronts . d) Approximate profiles of the heat wave for $\gamma = 5$. The effect of γ is stronger and exhibited by convex shock wave profiles.

$$\delta = \sqrt{\frac{D_M}{\alpha_s} J_0^{3+p} e^{2\alpha_s(3+p)}} \sqrt{\frac{N(n,p)}{2(4_p)}} \tag{53}$$

The ratio x/δ provides the dimensionless ratio $x/\sqrt{(D_M/\alpha_s)\, J_0^{3+p}}$ because $\left(D_M J_0^{3+p}/\alpha_s\right)$ has a dimension $[m^2]$. The normalized approximate solution (54) is

$$\frac{T_s(x,t)}{J_0 e^{2\alpha_s t}} = \left(1 - \frac{\eta_M}{\sqrt{N(n,p)/2}}\right)^n \tag{54}$$

and generates a new similarity variable [27] η_M, namely

$$\eta_M = \frac{x}{\sqrt{\frac{D_M}{\alpha_s} J_0^{3+p} e^{2\alpha_s t(3+p)}}} \tag{55}$$

while the denominator of (54) defines numerically the penetration depth $\sqrt{N(n,p)/2(4+p)}$ for $x/\delta = 1$.

However, for $0 < \mu < 1$ the integration of (51) is not straightforward and we have to use that

$$D_t^{-\mu}\left(e^{at}\right) = \frac{a^{-\mu} e^{at}}{\Gamma(\mu)} \gamma(\mu, at) = E_t(\mu, at) \tag{56}$$

where $E_t(\mu, at)$ is the E_t unction (57) , while $\gamma(\mu, at)$ is the incomplete Gamma function

$$E_t(\mu, at) = \frac{1}{\Gamma(\mu)} \int_0^t z^{\mu-1} e^{az} dz \tag{57}$$

with the special case $E_t(0, at) = lim_{\mu \longrightarrow 0 E_t(\mu,at)} = e^{at}$. Moreover, $E_t(\mu, a)$ satisfies the recurrence relation

$$E_t(\mu, a) = aE_t(\mu+1, a) + \frac{t^\mu}{\Gamma(\mu+1)} \tag{58}$$

Hence, after these explanations we get from (51)

$$\delta^2 = \frac{D_M N(n,p) J_0^{4+p} a_M^{-\mu} e^{a_M t} \gamma(\mu, a_M t)}{J_0 e^{2a_s t}} = \frac{D_M N(n,p) J_0^{4+p} E_t(\mu, a_M)}{J_0 e^{2a_s t}} \tag{59}$$

where a_M denotes the numerical value of the product of rate constant a_s of the boundary condition and the non-linearity $(4+p)$ of the diffusion term in (50), that is

$$a_M = (2\alpha_s)(4+p) \tag{60}$$

Rearranging the solution (59) we get

$$\delta^2 = D_M N(n,p) J_0^{3+p} \frac{a_M^{-\mu} e^{a_M t} \gamma(\mu, a_M t)}{e^{2\alpha_s t}} = J_0^{3+p} D_M N(n,p) \frac{Et(\mu, a_M}{e^{2\alpha_s t}} \tag{61}$$

With the second version of (56) and the definition (60) we obtain a more readable form, namely

$$\delta^2 = \frac{\left[J_0 e^{2\alpha_s t}\right]^{(3+p)}}{a_M^{\mu}} N(n,p)\gamma(\mu, a_M t) \tag{62}$$

Therefore, the penetration depth is

$$\delta = \sqrt{\frac{D_M}{a_M^{\mu}}} \left[J_0 e^{2\alpha_s t}\right]^{\frac{3+p}{2}} \sqrt{N(n,p)\gamma(\mu, a_M t)} \tag{63}$$

Here $\sqrt{\frac{D_M}{a_M^{\mu}} J_0^{3+p}}$ has a dimension $[m]$ and consequently the ratio x/δ defines a similarity variable

$$\eta_M^{\mu} = \frac{x}{\sqrt{\frac{D_M}{a_M^{\mu}} \left[J_0 exp(2\alpha_s t)\right]^{3+p}}} \tag{64}$$

For $\mu = 1$ we recover the similarity variable (55)

Then, the approximate solution is

$$T_a = J_0 exp(2\alpha_s t) \left(1 - \frac{\eta_M^{\mu}}{N_p^{\mu}}\right)^n \tag{65}$$

$$N_p^{\mu} = \sqrt{\frac{N(n,p)\gamma(\mu, a_M)}{[2(4+p)]^{\mu}}} = \sqrt{\frac{N(n,p)}{2(4+p)}} \sqrt{\gamma(\mu, a_M t)\left[2(4+p)\right]^{1-\mu}} \tag{66}$$

The normalized form of (65) is

$$\frac{T_a}{J_0 exp(2\alpha_s t)} = \left(1 - \frac{x}{\sqrt{\frac{D_M}{\alpha_s^{\mu}}\left[J_0 exp(2\alpha_s t)\right]^{3+p} N_p^{\mu}}}\right)^n \tag{67}$$

The condition $(\eta_M^{\mu}/N_p^{\mu}) = 1$ defines the front of the thermal wave, namely

$$x_{\delta,\mu} = \sqrt{\frac{D_M}{\alpha_s^{\mu}}\left[J_0 exp(2\alpha_s t)\right]^{3+p}} \tag{68}$$

The ratio $\frac{x_\delta}{x_{\delta,\mu}} = \frac{\delta}{\delta_\mu}$ allows to define the retardation factor $R_{\delta,\mu}$ due the memory effect, namely

$$R_{\delta,\mu} = \frac{\delta}{\delta_\mu} = \sqrt{\frac{\alpha_s}{a_N^{\mu}}}\sqrt{2(4+p)\gamma(\mu, a_M t)} \tag{69}$$

Taking into account the relationship (60) we get

$$R_{\delta,\mu} = \sqrt{\alpha_s^{1-\mu}}\sqrt{\left[2(4+p)\right]^{1-\mu}}\sqrt{\gamma(\mu, a_M t)} \tag{70}$$

For $\mu = 1$ we get $R_{\delta,\mu=1} = 1$. Moreover, the retardation factor $R_{\delta,\mu}$ is independent of the exponent n of the assumed profile and the integration technique used (DIM in this case). The only effect is due the exponent $(1-\mu)/2$ and it is related to the expected subdiffusion transport of the thermal wave.

As it was demonstrated in [27] the optimal exponent, for $\mu = 1$ and the restrictions imposed by the conditions in the problem of Marshak lead to the following limitation $1/8 < n_{opnopt} < 1/4$ about the optimal exponent. Further, with the DIM solution the minimization of $E(z,t)$ provided $n_{opt} \approx 0.153$ with $E(z,t) \approx 0.00245$, while with $p = 3$ used by Marshak the optimal exponent is $n_{opt} \approx 0.166$. Now, bearing in mind that for the degenerate diffusion equation, as it was demonstrated in the previous example, the with optimal exponents close to the value $n = 1/(p+4)$ (see eq. (49), we have $T^{4+p} \approx T^5$; where $p = 1$ is the values chosen by Marshak. With $\beta = 1$ in (49), which is equivalent to (27), we have $w = (4+\alpha)/\beta \approx 5.5$ and therefore the scaling is $T^w \approx T^{5.5}$. From the data summarized in Table 1 we have to expect that $n_{opt} \leq 1/5.5 \approx 0.18$. The minimization of the residual function yields values of the optimal exponent satisfying this restriction (see Table 3).

The profiles in Figure 3 are represented as functions of the dimensionless space coordinate $X = x\delta_\mu/$ which allows better to compare these solutions to the solutions developed for the case of $\mu = 1$ developed in [27].

Table 3. Optimal exponents of the approximate profile: Marshak's problem

$(4+p), \mu \rightarrow$	0.1	0.2	0.3	0.4	0.5	0.6	0.7	0.8	0.9	0.99
5.5	0.08	0.0802	0.074	0.068	0.061	0.052	0.040	0.030	0.015	0.012

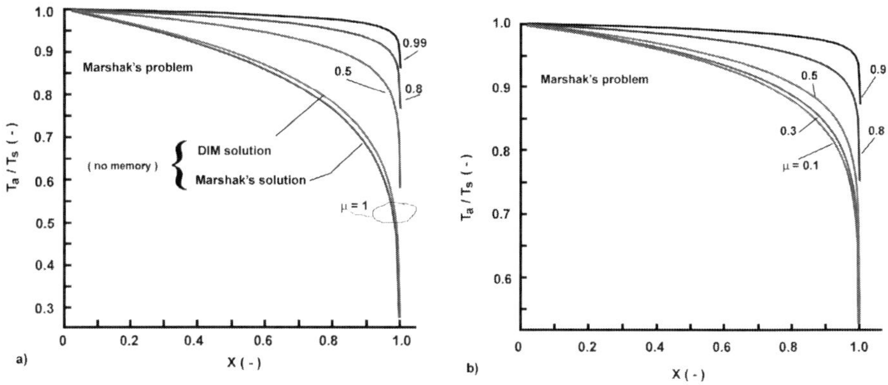

Figure 3. Marshak problem. Approximate heat wave profiles developed by DIM and expressed as functions of the variable $X = x/\delta_\mu$. Strong effects of the nonlinearity in the thermal diffusivity term $T^{5.5}$) and suppressed memory (subdiffusion) effects. a) Approximate normalized profiles of the heat wave for $\mu = 1$ and $\mu = 0.5$, $\mu = 0.8$, $\mu = 0.9$. b) Approximate normalized profiles of the heat wave for $\mu = 0.1$,$\mu = 0.3$, $\mu > 0.5$ and $\mu = 0.9$. The fronts of the profiles are steep as in the temperature distributions in Figure 2.

4.2.2. Garrnier's Problem: Power-Law Time Dependent Boundary Temperature

Garnier et al. [9] postulated, by change of variable $\zeta = T^{4+\alpha}$, the following integer-order model

$$\frac{\partial \xi^\epsilon}{\partial t} = D_G \frac{\partial^2 \xi}{\partial x^2}, \quad D_G = \frac{16}{12+3\alpha} \frac{g\sigma}{\rho_0^{(2-\mu_0+\lambda)}}, \quad \epsilon = \frac{\beta}{4+\alpha} \tag{71}$$

using the scaling laws (2) with $\alpha = 1.5$,$\beta = 1.6$, $\lambda = 0.2$ and $\mu_0 = 0.14$.

The solution of Garnier et al. [9] requires $2\beta < 4 + \alpha$, thus assuring $\epsilon \in (0, 1/2)$. Moreover, Garnier et al. [9] assumed the finite depth concept in the heat wave propagation imposing the condition $\xi(\delta) = \partial\xi(\delta)/\partial x = 0$, that is exactly the same as in the integral-balance approach used in this chapter.

Further, with a ramp heating expressed by a power-law surface temperature $T_s(t) = T_{s0}(t/t_s)^k$, the validity of the solution Garnier et al. [9] needs the condition $-1/(4+\alpha-\beta) < k < -1/(4+\alpha-2\beta)$ to be obeyed. In the definition of the boundary condition k is a given exponent and t_s is a time scale. The Garnier's approach resulted in a self-similar solutions with fronts moving as $\delta = \delta_0(t/t_s)^{n_G}$; where $\delta_0 = \sqrt{GD_G}\sqrt{T_{s0}^{4+\alpha}}$ and G is a dimensionless parameter (see [27] for more details).

With the concept of memory in the heat flux propagation, as it was done in section 2.43, we have a fractional analog of (70), namely

$$\frac{\partial^\mu \xi^\epsilon}{\partial t^\mu} = D_G \frac{\partial^2 \xi}{\partial x^2} \tag{72}$$

By change of the variable as $\theta = \xi^\epsilon = T^{4+\alpha}$ we get

$$\frac{\partial^\mu \theta}{\partial t^\mu} = D_G \frac{\partial \theta^w}{\partial x^2}, \quad w = \frac{4+\alpha}{\beta} = \frac{1}{\epsilon} \tag{73}$$

Now, assuming the approximate profile as $\theta = \theta_s(1 - x/\delta)^n$ and the DIM technology we have

$$\frac{1}{(n+1)(n+2)} \frac{\partial^\mu}{\partial t \mu} \delta^2 = D_G \frac{T_{s0}^{w-1}}{\alpha_{ts}} \tag{74}$$

or simply

$$\delta^2 = D_G N(n, w, \mu) \frac{T_{s0}^{w-1}}{\alpha_{ts}} t^{k(w-1)-\mu} \quad \alpha_{ts} = (t_s)^{kw-1} = const. \tag{75}$$

where

$$N(n, w, \mu) = \frac{(n+1)(n+2)}{w} \frac{\Gamma(k(w-1)+1)}{\Gamma(k(w-1)-\mu+1}\tag{76}$$

Hence, the penetration depth is

$$\delta_{G,\mu} = \sqrt{D_G T_{s0}^{(w-1)} t^{n_H}} \sqrt{\frac{N(n, w, \mu)}{a_{ts}}}, \quad n_H = k(w-1) - \mu \tag{77}$$

With the prescribed values of α and β we have $w \approx 3.437$ and $w-1 = 2.437$. Therefore, the front propagates proportional to $\delta \propto t^{2.437k-\mu}$. To have a subdiffusion behaviour, we need $(2.437k - \mu)/2 < 1$ that impose the condition $k < \frac{2+\mu}{2.437}$ or $\mu > 0.437$. Because, we have always $2+\mu < 3$, then the subdiffusion behaviour of the front motion requires $k < 1.23$. We have to remember, that in solids we have only subdiffusion transport, and this imposes the condition on the exponent in the time-depended (power-law) boundary conditions, namely: $2/2.437 = 0.820 < k < 1.231$. The condition of Garnier, mentioned above, means $-0.256 < k < 0.434$ for $\mu = 1$, but the negative sign of k implies a decaying ramp boundary condition that contradicts the physical conditions of the problem.

Next, from the solution developed and $\mu = 1$ we have $\delta \equiv t^{\frac{n_H}{2}}$ and $(n_H - 1) = (n_G - 1)/\beta$ as it was developed in [27]. Further, with the inverse transforms $T = \theta^{1/(4+\alpha)}$ the approximate solution $T_a(x,t)$ is

$$T_a(x,t) = T_{s0}\left(\frac{t}{t_s}\right)^k \left(1 - \frac{x}{\delta_{G,\mu}}\right)^q, \quad q = \frac{n}{4+\alpha} \tag{78}$$

The optimization of the residual function (with respect to q) provides $m = 2.347$. Then, we get $q_{opt(m=2.437)} \approx 0.636$ and consequently $n_{opt(m=2.437)} \approx 3.501$ [27]. Therefore, the normalized temperature profile is

$$\frac{T_a(x,t)}{T_{s0}\left(\frac{t}{t_s}\right)^k} = \left(1 - \frac{\eta_H}{(2.684/a_s(1+4.347k))}\right)^q, \quad q = \frac{n}{4+\alpha} \tag{79}$$

Here $n_{H(\mu=1)} = x/\sqrt{D_G T_{s0}^{n_H/2} t^{n_H}}$ is the similarity variable for $\mu = 1$, while for $0, \mu < 1$ the similarity variable is $n_{H(\mu\neq 1)} = x/\sqrt{D_G T_{s0}^{w-1} t^{n_H}}$

Consequently, the flux approximate profile is

$$\Phi_a(x,t) = D_G\left[T_{s0}\left(\frac{t}{t_s}\right)\right]^4 \frac{4q}{\delta_{G,\mu}}\left(1 - \frac{x}{\delta_{G,\mu}}\right)^{4q-1} \tag{80}$$

or normalized by its surface density as

$$\frac{\Phi_a(x,t)}{\left[T_{s0}\left(\frac{t}{t_s}\right)\right]^4}\frac{\delta_{G,\mu}}{4q} \equiv \left(1 - \frac{x}{\delta_{G,\mu}}\right)^{4q-1} \tag{81}$$

For $0 < \mu < 1$, the optimal exponents should be established by a procedure similar to that used in the previous examples. We skip this step since real values of the exponent k are missing in the literature and the determination of the optimal exponent would be a serious obstacle. Meanwhile, we may use the values $q_{opt(m=2.437)} \approx 0.636$ and $n_{opt(m=2.437)} \approx 3.501$ [27] to demonstrate what are deviations of the profiles (with respect to the cas with $\mu = 1$) when $0 < \mu < 1$. For the sake of simplicity of calculations we assume $\alpha_{ts} = 1$.

First of all, we are interested in the penetration depth evolution with effects of retardation controlled by the fractional parameter μ. The scaled penetration depth δ/δ_s, where $\delta_s = \sqrt{D_G T_{s0}^{(w-1)} t^{n_H}}$ is the length scale of the process presented as a function of the fractional parameter μ and the exponent k in Figure 4. The exponent k is within the range $0.820 < k < 1.231$, corresponding to subdiffusion penetration of the thermal wave, as it was established above. Besides, as it could be expected, the reduction in the fractional order μ results in sorter penetration depths, while the slower rise of the surface temperature (lower values of k) has an opposite effect.

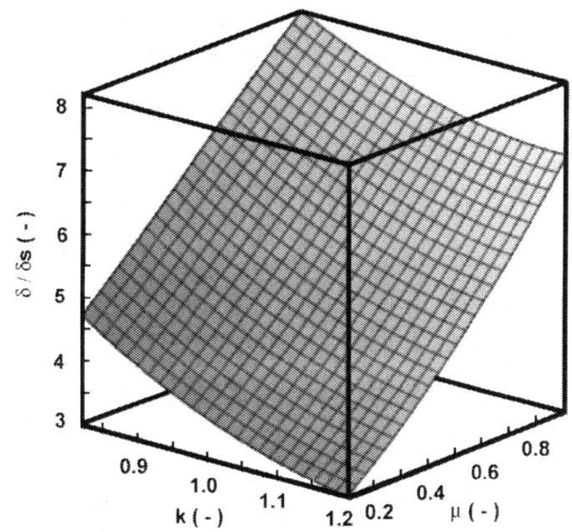

Figure 4. Scaled penetration depth as a function of the fractional order μ and the exponent k. $n = 3.437$.

Further, the three-dimensional profiles, that is, the surfaces of the thermal waves, are presented in Figure 5 as a function of the simultaneous action of the fractional order μ (the retardation effect of the medium) and the similarity variable η_H; while rate of the surface temperature rise through the exponent k

is taken as parameter. The two limits a) $k = 0.82$ and b) $k = 1.2$ are taken for these simulations corresponding to subdiffusion transport. The red line (in the electronic version) approximately defines the front of the wave. As we may expect, the reduction in μ results in shorter penetration depth.

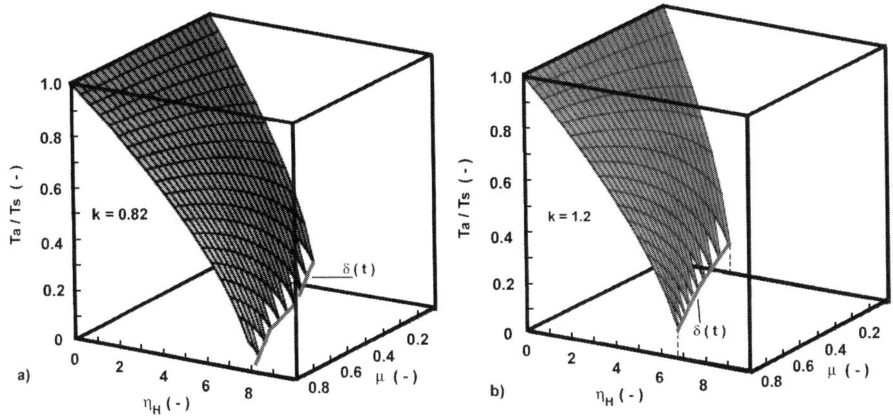

Figure 5. Garnier problem. Approximate three-dimensional heat wave profiles developed by DIM and expressed as functions of the similarity variable η_H and the fractional order μ. a) Case with $k = 0.82$ b) case with $k = 1.2$.

Conclusion

The chapter demonstrated development and approximate closed-form analytical solutions of the radiation heat diffusion with memory applying a constitutive approach involving a memory by a fractional integral with a singular (power-law) kernel. The solution approach utilized the double integration method allowing developments of closed-form solutions of the Dirichlet problem and two classical cases: Marshak's problems with exponential time-dependent boundary condition and Garnier's problem with power-law ramp heating of the surface. The solution approach is based on previously developed solution technology applied to integer-order degenerate parabolic equations [13] and non-linear subdiffusion equations [21]. It is generalization and development of the results obtained in [27].

References

[1] Rosseland S., *Astrophysik und atom-theretische Grundlagen.* Springer-Verlag, Berlin, 1931.

[2] Lonngren K. E., Ames W. F., Hirose A., Thomas J., Field penetration into plasma with nonlinear conductivity, *The Physics of Fluids* **17** , 1891–1920 (1974).

[3] Broadbridge P. , Lavrentiev M. M., Williams G. H., Non-linear heat conduction through externally heated radiant plasma: Background analysis for numerical study, *J. Math Anal. Appl.* **238**, 353-368 (1999).

[4] Tilloua A., Liberssart L., Joulin A., Monod B., Jeandel G., Determination of physical properties of fibrous thermal insulation, *EPJ Web of Conferences* **29**, 02009, doi:10.1051/epjconf/20123302009 (2012).

[5] Coleman B., Noll W., Foundations of linear Viscoelasticity, *Reviews of Modern Physics* **33**, 239–249 (1961).

[6] Smith C. C., Solutions of the radiation diffusion equation, *High Energy Density Physics* **6**, 48–56 (2010).

[7] Clourt J.-F., The Rosseland approximation for radiative transfer problems in heterogeneous media, *J. Quant.Specrosc. Radiat. Transfer* **58**, 33–43 (1997).

[8] Magyari E., Pantokratoras A., Note on the effect of thermal radiation in linearized Rosseland approximation on heat transfer in various boundary layer flows, *Int. Comm.Heat Mass Transfer* **5**, 99-108 (2011).

[9] Garnier J., Malinie G., Saillard Y., Cherfis-Clerouin C., Self-similar solution for non-linear radiation diffusion equation, *Phys. Plasmas 13*,doi:10.1063/1.2350167 (2006).

[10] Marshak R. E., Effect of Radiation on shock wave behaviour, *The Physics of Fluids* **1**, 24–29 (1958).

[11] Barenblatt G. I., On some unsteady motions of a liquid or a gas in a porous medium, *App. Math Mech.* **16**, 67–78 (1952).

[12] Zeldovitch Ya. B. and Raizer Yu. O., *Physics of shock waves and high temperature hydrodynamic phenomena* , Academic Press, New York, 1966.

[13] Hristov J., Integral solutions to transient nonlinear heat (mass) diffusion with a power-law diffusivity: a semi-infinite medium with fixed boundary conditions, *Heat Mass Transfer*, **52**, 635-655 (2016).

[14] Fabre A., Hristov J., On the integral-balance approach to the transient heat conduction with linearly temperature-dependent thermal diffusivity, *Heat Mass Transfer* **53**, 177-204 (2017).

[15] Hammer J. H. and Rosen M. D., A consistent approach to solving the radiation diffusion equation, *Phys. Plasmas Phys. Plasmas 10*,1829 doi: 10.1063/1.1564599 (2003).

[16] Hristov J., An Approximate Analytical (Integral-Balance) Solution to A Nonlinear Heat Diffusion Equation, *Thermal Science* **19**, 723-733 (2015).

[17] Boltzmann L., Zur Theorie der Elastischen Nachwirkung [Theory of elastic aftereffect], *Sitzungsber Akad. Wiss. Wien. Mathem.- Naturwiss.* **70**, 275–300 (1874).

[18] Hristov J., Transient Heat Diffusion with a Non-Singular Fading Memory: From the Cattaneo Constitutive Equation with Jeffreys kernel to the Caputo-Fabrizio time-fractional derivative, *Thermal Science*, **20**,765–770 (2016).

[19] Hristov J., Fractional derivative with non-singular kernels: From the Caputo-Fabrizio definition and beyond: Appraising analysis with emphasis on diffusion models, In: S. Bhalekar (Ed). *Frontiers in Fractional Calculus, 2017.* Bentham Science Publishers, Sharja, OAE, 2017, pp. 269-342.

[20] Podlubny I., *Fractional Differential Equations*, Academic Press, New York, 1999.

[21] Hristov J., Integral-balance solution to nonlinear subdiffusion equation, In: S. Bhalekar (Ed). *Frontiers in Fractional Calculus, 2017.* Bentham Science Publishers, Sharja, OAE, 2017, pp. 71-106.

[22] Goodman T. R., Application of Integral Methods to Transient Nonlinear Heat Transfer, In: Irvine TF and Hartnett JP (Eds.), *Advances in Heat Transfer*, **1** ,51-122, Academic Press, San Diego, CA, 1964.

[23] Hristov J., The heat-balance integral method by a parabolic profile with unspecified exponent: Analysis and Benchmark Exercises, *Thermal Science* **132**, 27–48 (2009).

[24] Hristov J., Approximate solutions to time-fractional models by integral balance approach, In: Cattani C., Srivastava H. M., Xia-Jun Yang, (Eds). *Fractional Dynamics.* De Gruyter Open, Warsaw/Berlin, 2015, pp. 78-109.

[25] Hill J. M., Similarity solutions for nonlinear diffusion- a new integration procedure. *J. Eng, Math.* **23**, 141-155 (2004).

[26] Prasad S. N., Salomon H. B., A new method for analytical solution of a degenerate diffusion equation,*Adv. Water Research*, **28**, 1091–1101 (2005).

[27] Hristov J., The heat radiation diffusion equation: Approximate explicit analytical solutions by improved integral-balance method, *Thermal Science*, **22**,777–788 (2018).

[28] Hristov J., Double Integral-Balance Method to the Fractional Subdiffusion Equation: Approximate solutions, optimization problems to be resolved and numerical simulations, *J. Vibration and Control* , **23**,2795–2818 (2017).

[29] Myers T., Optimizing the exponent in the hear balance and refined integral methods, *Int. Comm. Heat Mass Transfer*, **44**, 143–147 (2008).

[30] Mitchell S. L., Myers T. G., Application of standard and rened heat balance integral methods to one-dimensional Stefan problems, *SIAM Review*, **52**,57–86 (2010).

[31] Zener C., Theory of growth of spherical precipitates from solid solutions, *J Appl. Phys*, **20**, 950–953 (1949).

[32] Langford D., The heat balance integral method, *Int. J. Heat Mass Transfer*, **16**, 2424–2428 (1973).

In: Heat Conduction
Editors: J. Hristov and R. Bennacer
ISBN: 978-1-53614-673-8

Chapter 4

Fundamental Solutions to the Cauchy and Dirichlet Problems for a Heat Conduction Equation Equipped with the Caputo-Fabrizio Differentiation

***Derya Avci*[1,*], *Mehmet Yavuz*[2] *and Necati Özdemir*[1]**
[1]Department of Mathematics, Faculty of Sciences and Arts, Balikesir University, Balikesir, Turkey
[2]Department of Mathematics-Computer Sciences, Faculty of Science, Necmettin Erbakan University, Konya, Turkey

Abstract

In this chapter, a heat conduction equation in terms of the Caputo-Fabrizio derivative of order $0 < \alpha \leq 1$ is considered. Caputo-Fabrizio derivative is defined by a non-singular exponential decay kernel function. Thanks to this feature, it eliminates the computational difficulties arising from the singular power kernel functions of the traditional fractional derivatives such as Riemann-Liouville and Caputo. In the present study, the fundamental solutions to the Cauchy and Dirichlet problems based upon a heat conduction equation equipped with the Caputo-Fabrizio derivative are investigated on a line segment. To obtain the fundamental

*Corresponding Author Email: dkaradeniz@balikesir.edu.tr.

solutions, the Laplace transform with respect to time t and the finite sin-Fourier transform with respect to spatial coordinate x are applied. The temperature profiles are illustrated for different values of fractional order α by the graphics.

Keywords: Caputo-Fabrizio derivative, heat conduction equation, fundamental solution, Laplace transform, finite sin-Fourier transform

1. Introduction

The non-classical theories of heat conduction, in which the standard heat conduction equation has been replaced by more general equations in terms of fractional order time/space derivatives, have an increasing interest among the researchers [1, 2, 3]. Because the generalized heat models with fractional derivatives make it possible to identify the sub/super behaviours (anomalous structures) of the heat conduction process. In the literature, many studies related to the heat conduction equation in terms of conventional fractional derivatives such as Riemann-Liouville (RL) and Caputo have been studied successfully from different point of views. Povstenko has detailed the fundamental solutions and physical motivation of initial-boundary value problems based on fractional heat conduction equation in different coordinate systems [4].

In 2015, Caputo and Fabrizio have given a new perspective to fractional calculus by introducing a new operator without singular kernel [5]. By analogy, Atangana and Baleanu have proposed two types of fractional derivative in sense of RL and Caputo, based uponwith the generalized Mittag-Leffler function that can be more suitable for modelling of some diffusive transports in the nature than the power function [6]. On the other hand, some researchers have concluded that these new non-singular operators can be considered as a filter regulator in the engineering applications [7].

Caputo-Fabrizio (CF) and Atangana-Baleanu (AB) operators have been used to model the electromagnetic waves in dielectric media [8], to analyze the diffusive processes [9], to price the financial instruments [10], to model chaotic behaviours of a nonlinear systems [11], to show how a CF space-fractional derivative can be improved starting from an analogue of the Cattaneo relation with a space-dependent fading memory [12], to compare CF and AB fractional derivatives with Allen-Cahn model [13] and generalized Casson fluid model [14], to apply to groundwater flow within confined aquifer [15], to approach with numerical solutions from Riemann-Liouville definition to AB definition [16], to obtain the solutions of PDEs by using Laplace perturbation method [17], to compare CF and AB fractional derivatives with lineer/nonlinear PDEs [18], etc. Mirza and Vieru [19] have proposed the fundamental solutions to the advection-diffusion equation (ADE) with CF derivative in two dimensional space. Baleanu et al. [20] have researched the numerical solutions of ADE with the CF derivative and also have given a detailed comparison with the existing results for ADE with Caputo fractional derivative. Alkahtani and Atangana have solved the existence and uniqueness problems for a nonlinear heat problem in terms of CF derivative and they have also explained the filter role of CF derivative by using their nonlinear model [21].

The main aim of this study is to demonstrate the fundamental solutions of the Cauchy and the Dirichlet problems based on the heat conduction equation modelled with CF derivative. For this purpose, some basic definitions used throughout the formulation are given in Section 2 The Cauchy problem is solved by using the integral transforms and the results are illustrated by the graphics in Section 3 Likewise, the Dirichlet problem is analyzed in Section 4 In the figures, the temperature profile represented by the fundamental solutions are shown under the variation of fractional order. Finally, the concluding remarks are given.

2. Preliminaries

Let us remind some well-known mathematical descriptions used in the present study.

Definition 1. Let $a \in [-\infty, t)$, $f \in H^1(a,b)$ and $b > a$. The Caputo fractional derivative of order α is defined by [22]

$$^{C}D_t^{\alpha} f(t) = \frac{1}{\Gamma(n-\alpha)} \int_0^t (t-\tau)^{n-\alpha-1} \frac{d^n f(\tau)}{d\tau^n} d\tau, \; n-1 < \alpha \leq n, \;\; n \in \mathrm{N}. \tag{1}$$

Notice that Caputo fractional derivative finds numerous applications in the areas of science and engineering because of its well-equipped properties such as Laplace transform rule.

Definition 2. The Laplace transform of Caputo fractional derivative $^{C}D_t^{\alpha} f(t)$

has the form [22]:

$$\mathcal{L}\left\{{}^{C}D_t^{\alpha}f(t)\right\} = s^{\alpha}\mathcal{L}\{f(t)\} - \sum_{k=0}^{n-1} f^{k}\left(0^{+}\right)s^{\alpha-1-k}, \quad n-1<\alpha\leq n. \quad (2)$$

The main idea of CF derivative is based on the changing of the kernel $(t-\tau)^{-\alpha}$ with the function $\exp\left(-\frac{\alpha}{1-\alpha}t\right)$ and $\frac{1}{\Gamma(1-\alpha)}$ with $\frac{M(\alpha)}{\Gamma(1-\alpha)}$ where $M(\alpha)$ is a normalization function such that $M(0) = M(1) = 1$. Therefore, the Caputo-Fabrizio fractional derivative is defined as follows

$$^{CF}D_t^{\alpha}f(t) = \frac{M(\alpha)}{1-\alpha}\int_0^t f'(\tau)\exp\left[-\frac{\alpha(t-\tau)}{1-\alpha}\right]d\tau. \quad (3)$$

The CF definition can also be applied for the functions which do not belong to $H^1(a,b)$ and the kernel does not have singularity for $t=\tau$ unlike the Caputo fractional derivative. In the present study, we apply the Laplace transform of CF derivative with respect to time variable t. From the property of Laplace transform of convolution, The Laplace transform of $^{CF}D_t^{\alpha}f(t)$ is defined as follows [5, 23]:

$$\mathcal{L}\left\{{}^{CF}D_t^{\alpha+n}f(t)\right\}(s) = \frac{s^{n+1}\mathcal{L}\{f(t)\} - s^{n}f(0) - s^{n-1}f'(0) - \cdots - f^{(n)}(0)}{s+\alpha(1-s)}. \quad (4)$$

In addition, we use the finite sin-Fourier transform with respect to spatial coordinate x. In the finite interval $0\leq x\leq L$, the corresponding transform is given by [24]

$$\mathcal{F}\{u(x)\} = \widetilde{u}(\xi_k) = \int_0^L u(x)\sin(\xi_k x)\,dx, \quad (5)$$

with its inverse transform

$$\mathcal{F}^{-1}\{\widetilde{u}(\xi_k)\} = u(x) = \frac{2}{L}\sum_{k=1}^{\infty}\widetilde{u}(\xi_k)\sin(\xi_k x), \quad (6)$$

where $\xi_k = \frac{k\pi}{L}$, $k = 1,2,3,\ldots$. The finite sin-Fourier transform of the second order derivative of a given function is given by the following relation

$$\mathcal{F}\left\{\frac{d^2u(x)}{dx^2}\right\} = -\xi_k^2\widetilde{u}(\xi_k) + \xi_k\left[u(0) - (-1)^k u(L)\right]. \quad (7)$$

In the classical or fractional theory of partial differential equations, Dirac delta function is commonly assumed as initial or boundary condition function because of its convenience with the following integral transform properties:

$$\mathcal{L}\left\{\delta\left(t\right)\right\} = 1,$$
$$\mathcal{L}\left\{\delta\left(t-a\right)\right\} = e^{-as} \text{ for } a > 0.$$

Let $\{\varphi_n\left(x\right) : n = 0, 1, 2, ...\}$ be a class of real orthogonal functions which occur as solutions of ordinary differential equations of the Sturm-Liouville problems. They satisfy the orthogonality property on the interval (a, b) may be infinite at either end or both ends. In this case, Dirac delta function obeys the following property [25]

$$\int_a^b \delta\left(x-t\right)\varphi_n\left(x\right)dx = \varphi_n\left(t\right),\ t \in \left(a, b\right).$$

Moreover, if a function $F\left(t\right)$ can be expanded in a series of orthogonal functions $\varphi_n\left(x\right)$, the completeness property is given by

$$\int_a^b F\left(t\right)\delta\left(t-x\right)dt = F\left(x\right).$$

3. Fundamental Solution of The Cauchy Problem

Let us consider the heat conduction equation in terms of CF operator

$$\frac{\partial^\alpha T}{\partial t^\alpha} = a\frac{\partial^2 T}{\partial x^2},\quad 0 < x < L,\ t > 0, \tag{8}$$

with the initial condition

$$t = 0\ :\ T\left(x, 0\right) = \delta\left(x - x_0\right),\ 0 < x_0 < L, \tag{9}$$

and the zero boundary conditions

$$\begin{aligned} x = 0\ &:\ T\left(0, t\right) = 0, \\ x = L\ &:\ T\left(L, t\right) = 0, \end{aligned} \tag{10}$$

where a denotes the thermal diffusivity constant. For simplicity, we assume $a = 1$ throughout the paper. By applying the Laplace transform with respect to

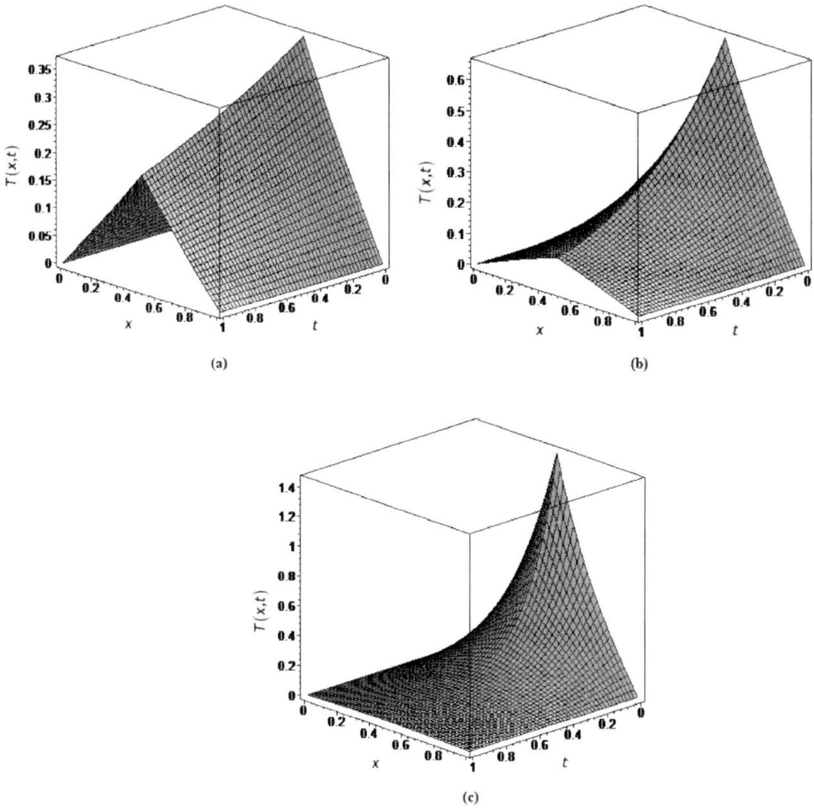

Figure 1. Fundamental solutions of the Cauchy problem for the values **(a)** $\alpha = 0.4$ **(b)** $\alpha = 0.7$ **(c)** $\alpha = 0.9$.

time variable t and finite sin-Fourier transform with respect to spatial coordinate x leads to

$$\frac{s\widetilde{T}^*(\xi_k, s) - \sin(\xi_k x_0)}{s + \alpha(1-s)} = -a\xi_k^2 \widetilde{T}^*(\xi_k, s).$$

After some mathematical arrangements, we get

$$\widetilde{T}^*(\xi_k, s) = \frac{\gamma \sin(\xi_k x_0)}{a\xi_k^2 + \gamma} \frac{1}{s + \frac{a\alpha\xi_k^2\gamma}{a\xi_k^2 + \gamma}}, \tag{11}$$

where $\gamma = \frac{1}{1-\alpha}$, $\xi_k = \frac{k\pi}{L}$, $k = 1, 2, 3, ...$

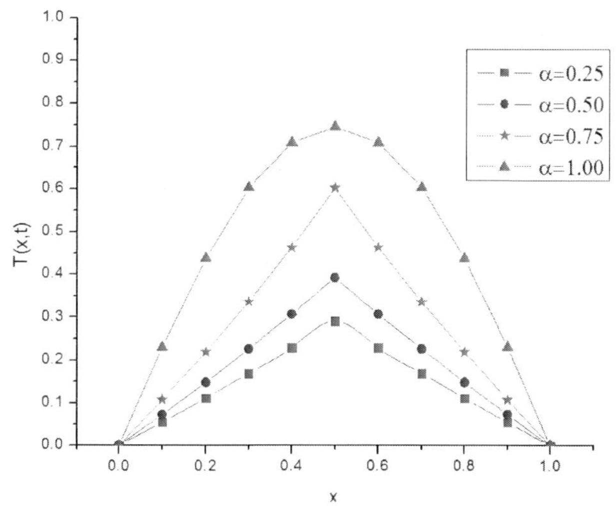

Figure 2. A Comparison on the fundamental solutions to the Cauchy problem for changing values of α.

Next, taking the inverse Laplace transform of Eq. (11), we arrive

$$\widetilde{T}\left(\xi_k, t\right) = \frac{\gamma \sin\left(\xi_k x_0\right)}{a\xi_k^2 + \gamma} e^{-\frac{a\alpha\xi_k^2\gamma}{a\xi_k^2+\gamma}t}. \tag{12}$$

After inverting the finite sin-Fourier transform in Eq. (12), we obtain the fundamental solution as

$$T\left(x, t\right) = \frac{2}{L}\sum_{k=1}^{\infty}\left[\frac{\gamma \sin\left(\xi_k x_0\right)}{a\xi_k^2 + \gamma} e^{-\frac{a\alpha\xi_k^2\gamma}{a\xi_k^2+\gamma}t}\right]\sin\left(\xi_k x\right), \tag{13}$$

In the limit case for $\alpha \to 1$, i.e. $\gamma \to \infty$, we derive the classical fundamental solution of the corresponding Cauchy problem,

$$T\left(x, t\right) = \frac{2}{L}\sum_{k=1}^{\infty}\sin\left(\xi_k x_0\right) e^{-at\xi_k^2}\sin\left(\xi_k x\right), \ x_0 \in (0, L). \tag{14}$$

The physical features with respect to the fundamental solution of Cauchy problem is illustrated by the Figure 1 and Figure 2. According to the figures, the results correspond to a slow heat conduction for decreasing values of α. $x_0 = 0.5$

is carried out in the Figure 1 and Figure 2. Notice that slow heat conduction process can also be modelled with Caputo fractional derivative. However, there are some generalized functions such as Mittag-Leffler, Mainardi, Fox-H and hypergeometric functions arising in the solutions of such problems which can be solved by introducing some numerical algorithms to remove the computational complexities. In this sense, CF derivative has a considerable advantage compared with Caputo fractional derivative.

4. Fundamental Solutions to the Dirichlet Problem

In this section, we consider the following Dirichlet problem:

$$\frac{\partial^\alpha T}{\partial t^\alpha} = a\frac{\partial^2 T}{\partial x^2}, \quad 0 < t < \infty, \tag{15}$$

with the homogeneous initial condition

$$t = 0 \ : \ T(x, 0) = 0,$$

and the boundary conditions

$$\begin{aligned} x = 0 \ &: \ T(0, t) = \delta(t), \\ x = L \ &: \ T(L, t) = 0. \end{aligned} \tag{16}$$

Note that we assume the physical quantities like thermal diffusivity parameter a as non-dimensional arbitrary constants only for the convenience.

Now, we apply the integral transforms similar to Section 3 and then have the following result

$$\frac{s\widetilde{T}^*(\xi_k, s)}{s + \alpha(1 - s)} = -a\xi_k^2\widetilde{T}^*(\xi_k, s) + a\xi_k$$

or equivalently,

$$\widetilde{T}^*(\xi_k, s) = \frac{a\xi_k}{a\xi_k^2 + \gamma} + \frac{a\alpha\xi_k\gamma^2}{\left(a\xi_k^2 + \gamma\right)^2}\frac{1}{s + \frac{a\alpha\xi_k^2\gamma}{a\xi_k^2+\gamma}}. \tag{17}$$

By inverting the integral transforms, we have the fundamental solution as follows

$$T(x, t) = \frac{2}{L}\sum_{k=1}^{\infty}\left[\frac{a\delta(t)\xi_k}{a\xi_k^2 + \gamma} + \frac{a\alpha\xi_k\gamma^2}{\left(a\xi_k^2 + \gamma\right)^2}e^{-\frac{a\alpha\xi_k^2\gamma}{a\xi_k^2+\gamma}t}\right]\sin(\xi_k x), \tag{18}$$

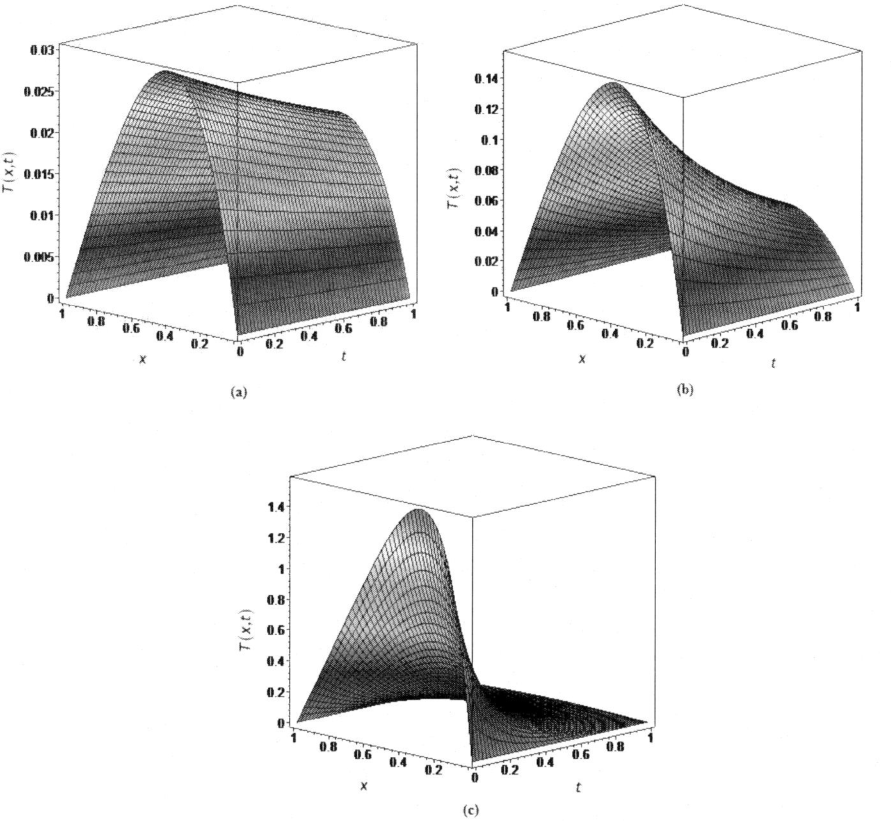

Figure 3. Fundamental solutions to the Dirichlet problem for the values **(a)** $\alpha = 0.3$ **(b)** $\alpha = 0.6$ **(c)** $\alpha = 0.9$.

Notice that $\delta(t) = 0$ when $t > 0$. This term is a characteristic result of the CF derivative as seen in [19]. The classical solution of the Dirichlet problem in the limit case $\alpha \to 1$, i.e., $\gamma \to \infty$, can be given as

$$T(x,t) = \frac{2a}{L} \sum_{k=1}^{\infty} \xi_k e^{-a\xi_k^2 t} \sin(\xi_k x). \tag{19}$$

Fundamental solution of the Dirichlet problem described by the Eqs. (15)-(16) is shown with the Figure 3 and Figure 4. These figures address the temperature profiles for different values of α.

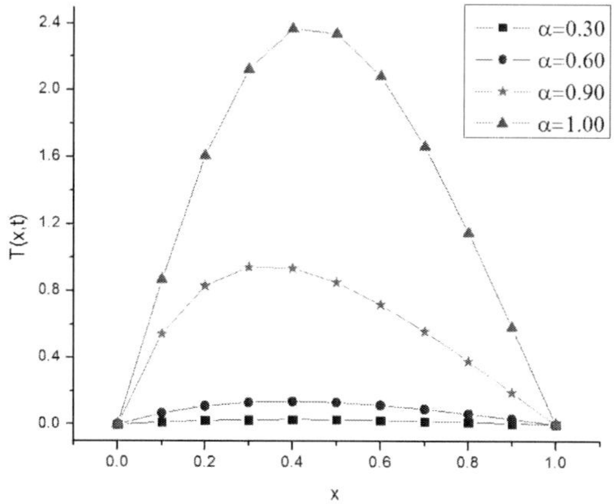

Figure 4. A Comparison on fundamental solutions to the Dirichlet problem for changing values of α.

Conclusion

In this chapter, we have considered the heat conduction equation with Caputo-Fabrizio operator in a line segment $0 < x < L$. The fundamental solutions for the Cauchy and the Dirichlet problems have been investigated by applying the integral transform techniques. The results have been illustrated by the figures under the variation of fractional order α. We have noticed that CF operator is a powerful tool to describe the sub/super structures in anomalous temperature distribution. In addition, the fundamental solutions have been obtained in terms of elementary functions such as exponential and trigonometric. This is a significant mathematical difference between the CF and Caputo models for heat diffusion. The Laplace transform of CF derivative needs some physically interpretable initial conditions and so it leads to computational convenience. Therefore, one can easily achieve the analytical solutions as seen in the present work. In the future studies under consideration, the thermal stresses due to heating process modelled with CF derivative.

References

[1] Povstenko Y., Fractional Heat Conduction Equation and Associated Thermal Stress, *Journal of Therm. Stres.,* 28, 83-102 (2005).

[2] Jiang X., Xu M., The time fractional heat conduction equation in the general orthogonal curvilinear coordinate and the cylindrical coordinate systems, *Physica A: Stat. Mec. Its Appl.,* 389, 3368-3374 (2010).

[3] Atanacković T., Konjik S., Oparnica L., Zorica D., The Cattaneo type space–time fractional heat conduction equation, *Contin. Mech. Thermodyn.,* 24, 293-311 (2012).

[4] Povstenko Y., *Fractional Thermoelasticity,* Springer, New York, (2015).

[5] Caputo M., Fabrizio M., A New Definition of Fractional Derivative without Singular Kernel, *Progr. Fract. Differ. Appl.,* 1, 1-13 (2015).

[6] Atangana A., Baleanu D., New Fractional Derivatives with Nonlocal and Non-Singular Kernel: Theory and Application to Heat Transfer Model, *Therm. Sci.,* 20, 763-769 (2016).

[7] Atangana A., Alkahtani B. S. T., New Model of Groundwater Flowing within a Confine Aquifer: Application of Caputo-Fabrizio Derivative, *Arab. J. Geosci.,* 9, 8 (2016).

[8] Gómez-Aguilar J., Escobar-Jiménez R., López-López M., Alvarado-Martínez V., Atangana-Baleanu Fractional Derivative Applied to Electromagnetic Waves in Dielectric Media, *J. Elect Waves and Appl.,* 30, 1937-1952 (2016).

[9] Hristov J., Derivatives with Non-Singular Kernels from the Caputo-Fabrizio Definition and Beyond: Appraising Analysis with Emphasis on Diffusion Models, in: *Frontiers in Fractional Calculus,* (Ed. S. Bhalekar), pp. 270-342 (2017).

[10] Yavuz M., Özdemir N., European Vanilla Option Pricing Model of Fractional Order without Singular Kernel, *Fractal and Fract.,* 2, 3 (2018).

[11] Atangana A., Koca I., Chaos in a Simple Nonlinear System with Atangana–Baleanu Derivatives with Fractional Order, *Chaos, Sol. Fract.,* 89, 447-454 (2016).

[12] Hristov J., Steady-State Heat Conduction in a Medium with Spatial Non-Singular Fading Memory: Derivation of Caputo-Fabrizio Space-Fractional Derivative with Jeffrey's Kernel and Analytical Solutions, *Therm. Sci.,* 21, 827-839 (2017).

[13] Algahtani O. J. J., Comparing the Atangana–Baleanu and Caputo–Fabrizio Derivative with Fractional Order: Allen Cahn Model, *Chaos, Sol. Fract.,* 89, 552-559 (2016).

[14] Sheikh N. A., Ali F., Saqib M., Khan I., Jan S. A. A., Alshomrani A. S., Alghamdi M. S., Comparison and Analysis of the Atangana–Baleanu and Caputo–Fabrizio Fractional Derivatives for Generalized Casson Fluid Model with Heat Generation and Chemical Reaction, *Res. in Phys.,* 7, 789-800 (2017).

[15] Atangana A., Baleanu D., Caputo-Fabrizio Derivative Applied to Groundwater Flow within Confined Aquifer, *J. Eng. Mech.,* 143, D4016005-1-5 (2017).

[16] Atangana A., Gómez-Aguilar J. F., Numerical Approximation of Riemann-Liouville Definition of Fractional Derivative: From Riemann-Liouville to Atangana-Baleanu, *Num. Met. Part. Diff. Eq.,* 33, 1-22 (2017).

[17] Yavuz M., Ozdemir N., Baskonus H. M., Solutions of Partial Differential Equations Using the Fractional Operator Involving Mittag-Leffler Kernel, *Eur. Phys. J. Plus,* 133, 215 (2018)

[18] Yavuz M., Ozdemir N., Comparing the New Fractional Derivative Operators Involving Exponential and Mittag-Leffler Kernel, *Dis. Cont. Dyn. Sys. Series S,* 12, 703-714 (2019).

[19] Mirza I. A., Vieru D., Fundamental Solutions to Advection–Diffusion Equation with Time-Fractional Caputo–Fabrizio Derivative, *Comp. Math. with Appl.,* 73, 1-10 (2017).

[20] Baleanu D., Agheli B., Al Qurashi M. M., Fractional Advection differential Equation within Caputo and Caputo-Fabrizio derivatives, *Advances in Mechanical Engineering,* 8, 1-8 (2016).

[21] Alkahtani B. S. T., Atangana A., Analysis of non-homogeneous heat model with new trend of derivative with fractional order, *Chaos, Sol. Fract.,* 89, 566-571 (2016).

[22] Kilbas A. A., Srivastava H. M., Trujillo J. J., Theory and applications of fractional differential equations. *North-Holland Mathematics Studies*, vol. 204, Elsevier, Amsterdam, (2006).

[23] Escamilla A. C., Aguilar J. F. G., Baleanu D., Escobar-Jimenez R. F., Olivares-Peregrino V. H., Abundez-Pliego A., Formulation of Euler-Lagrange and Hamilton equations involving fractional operators with regular kernel, *Adv. Differ. Equ.,* 2016, 283 (2016).

[24] Povstenko Y., *Linear fractional diffusion-wave equation for scientists and engineers.* Birkhauser, New York, (2015).

[25] Arfken G. B., Weber H. J., Harris F. E., *Mathematical Methods for Physicists: A Comprehensive Guide,* Academic Press, (2013).

In: Heat Conduction
Editors: J. Hristov and R. Bennacer
ISBN: 978-1-53614-673-8

Chapter 5

ABOUT THE HEAT CONDUCTION IMPACT ON THE THERMOSOLUTAL STABILITY WITHIN ANNULUS: SILICON CARBIDE CERAMIC CASE

Karim Ragui[1,*]***, Youb Khaled Benkahla***[1] ***and Rachid Bennacer***[2]
[1]Laboratory of Transfer Phenomenon, USTHB, Algiers, Algeria
[2]LMTENS-Paris-Saclay CNRS/Université Paris Saclay, Cachan, France

Abstract

This chapter summarizes some recent results of the conduction impact within a double diffusive convection of a viscoplastic fluid. Such a phenomenon was carried out within a Silicon-Carbide Ceramic annulus, which surrounded by a cold (and less concentric) outer cylinder and a hot (and concentric) inner one. The prediction of such a critical limit was a function of some relevant parameters as the fluid character, the boundary layers ratio, as well as the thermosolutal buoyancy one, to name but a few. As the impact of the geometry aspect ratio was taken into consideration, a noising flow phenomenon which called oscillation was inspected through this paper.

Keywords: conduction limits , thermosolutal convection, viscoplastic materials, porous annulus, circular shape, cut-cell approach, finite volume approach.

*Corresponding Author Email: ragui-karim@live.fr.

1. Introduction

1.1. Thermosolutal Convection Over the Years

During decades, the field of heat and mass transfer within a porous medium has been the subject of a very intense research activity, due to its importance of related industries and contemporary technological applications. Prominent among these applications being geothermal energy resources, nuclear energy systems, petroleum reservoirs, analysis of insulating systems, grain storage, pollutant dispersion in aquifers, processes of crystal manufacture, foam metals and fibrous media, to name but a few [1–3].

With both temperature and concentration gradients present to drive the fluid flow, an increased number of transfer areas was achievable with parallel or perpendicular gradients [4–10]. In the main idea to predict the transfer coefficients within such areas, powerful expressions have been developed over the years. Cited the work of Trevisan and Bejan [11] that was carried out in 1987, Lin et al. [12] in 1990, Bennacer [13] in 1993 and Ragui et al. [14], [15] recently, to name but a few.

Regarding its complexity in the industry, the flow of Newtonian fluids could present a simple limit of Non-Newtonian ones, such as crude oil and pharmaceutical products. The illumination of Bingham's researches, which carried out between 1916 and 1922 [16, 17], brought to light a new research axis for such complex fluids, called after as *Bingham plastics*. Such viscoplastic materials could act as a Newtonian fluid under a specific shear stress value and a gel-like material up to that value. Because of that, the prediction of the transfer rates of such fluids was found to be very complicated, experimentally speaking or numerically [18].

Within a porous matrix, viscoplastic materials have reached another level of complexity [19, 20]. Over the years, the permeability as well as the porosity of the porous matrix were the key factor in defining the behavior the yielded and the unyielded regions. Such a key factor produced some novel approaches which made the porous-diameter distribution within an unknown material very easy to be predicted [21].Nevertheless, only the thermal buoyancy impact was identified at that moment.

Side by side with the space shape, the thermal gradient impact has received a significant attention, especially when we talk about a square enclosure [22–24]. In fact, the restart and the control of the Bingham flow still a major issue for oil companies, due to the complex relationship between the shape of its industrial devices and the relevant parameters of the transfer phenomenon [25, 26].

1.2. About the Heat Conduction Limit

Through many years, thermal buoyancy has been treated as a key factor to describe the limit of the thermal behavior within various areas and fluids. Such a limit, which is a pure function of Rayleigh and Prandtl values, was defined in works of Kamotani [27], Travison and Bejan [28] and Bennacer [29] in which, a dominant conduction mode against a solutal one is described in complex abacuses.

The main question was about the behavior of the boundary layers and the relationship between them to make clear the pure conduction range and the mass diffusion as well. Looking for an answer, Bennacer [29] proved that the distinction between a heat transfer driven flow and a mass driven one depends on several criteria. These latter could be summarized by the ratio $(N/Le^{\alpha})^{\beta}$. Subsequently, the scale on which the buoyancy forces could act on the fluid flow is of the same importance as the relative intensity of the thermal and solutal forces. Such a complex conclusion has been numerically verified using Newtonian fluids.

1.3. Aim

Due to the fact that the conduction-limit impact within a double diffusive convection of viscoplastic materials is less frequent in the available literature, our chapter will discuss predictions of this phenomenon within a porous annulus layer. The solid matrix will be bounded by two isothermal and iso-concentric coaxial cylinders.

Complement to what was already presented with Newtonian fluids, the main purpose is to highlight the impact of some relevant parameters on the boundary layers of such a complex fluid, and to draw exhaustive conclusions on the transfer regimes of this kind of fluids. Such a studied geometry could plot an interesting work for many industrial applications, especially in refining and petrochemical fields.

2. Conductive Fluids between Modelling and Constitutive Equations

As known over the years, conductive fluids in industrial processes (*such as refining and petrochemical processes, as well as polymer melts industry*), can be divided between Newtonians and Non-Newtonian ones [30]. As a simple description of the first family, i.e., *the Newtonian fluids*, we bring to light air, water, ethylene glycol, and some chemical solutions [30–33] to name but a few. Since the 1990s, such a class of fluids is denoted as a limit case of some Non-Newtonian ones [16, 17, 34, 35].

As simple as we can abridge, and interested by the most adopted part of such complex fluids within the modern industry, viscoplastic materials have appeared over the years as various categories. Noted the Bingham in 1922 [16, 17], the Herschel-Bulkley in 1926 [34] and Casson 1959 [35]. Since then, several experimental and numerical predictions of such fluids have been developed in order to clarify its rheological-flow complexity and why its conduction regime can be easily achieved [36–40]. Typical examples of such fundamental works could be found in the recent paper of Mitsoulis Tsamopoulo [18] and the references therein.

Back in time, and between 1922 and 1987, modelling the stress-deformation behavior of viscoplastic materials was a very interesting question. Consequently, various constitutive equations have been proposed [30]. As a simple shear flow, the latter can take the following form:

$$\begin{cases} \tau^{k*} = {\tau_0}^{k*} + A\dot{\gamma}^n & |\tau| > \tau_0 \\ \dot{\gamma} = 0 & |\tau| \leq \tau_0 \end{cases} \tag{1}$$

where τ is the shear stress, $\dot{\gamma}$ is the shear rate. τ_0 is denoted as the yield stress.

In the above, the constant value A could be to the constant plastic viscosity μ_p or $\mu_p^{1/2}$ when we talk about Bingham or Casson models, respectively, or the consistency index K^* with the Herschel-Bulkley one. Similarly, n is the power law index which can be equal to unity $(n = 1)$ for the Bingham model or a half $(n = 0.5)$ for the Casson one. Unlike Bingham and Herschel-Bulkley fluids, where the constant k^* is equal to unity $(k^* = 1)$, the Casson model is defined with $(k^* = 0.5)$. Note that when the shear stress τ falls below τ_0, a solid structure is formed. Subsequently, an *unyielded* phenomenon comes to light.

In order to express this definition as a fully invariant constitutive relation (relevant in 3D), the tensors are introduced [41], [42]. Subsequently, the above model is written as follows:

$$\begin{cases} \overline{\overline{\tau}} = \left(A|\dot{\gamma}|^{n-1} + \frac{{\tau_0}^{k*}}{|\dot{\gamma}|^{k*}} \right) \overline{\overline{\dot{\gamma}}} & |\tau| > \tau_0 \\ \overline{\overline{\dot{\gamma}}} = 0 & |\tau| \leq \tau_0 \end{cases} \tag{2}$$

where $|\dot{\gamma}|$ is the magnitude of the symmetric rate-of strain tensor $\left|\overline{\overline{\dot{\gamma}}}\right|$, which can be defined as:

$$\begin{cases} |\dot{\gamma}| = \left(\frac{1}{2} \prod_{\dot{\gamma}} \right)^{1/2} = \left[\frac{1}{2} \left\{ \overline{\overline{\dot{\gamma}}} : \overline{\overline{\dot{\gamma}}} \right\} \right]^{1/2} \\ \overline{\overline{\dot{\gamma}}} = \nabla \overline{v} + \nabla \overline{v}^T \end{cases} \tag{3}$$

$\prod_{\dot{\gamma}}$ is the second invariant of $\overline{\overline{\dot{\gamma}}}$, $\nabla \overline{v}$ is the velocity gradient tensor; when $\nabla \overline{v}^T$ is its transpose.

In same way, $|\tau|$ is the magnitude of the extra stress tensor $\overline{\overline{\tau}}$:

$$|\tau| = \left(\frac{1}{2} \prod_{\tau} \right)^{1/2} = \left[\frac{1}{2} \left\{ \overline{\overline{\tau}} : \overline{\overline{\tau}} \right\} \right]^{1/2} \tag{4}$$

where $\prod_{\tau}$ is the second invariant of the extra stress tensor $\overline{\overline{\tau}}$.

Following the last presentation, the criterion to track down the yielded and the unyielded regions of such complex fluids is for the material to flow (yield) only when the magnitude of the extra stress tensor $|\tau|$ exceeds the yield stress τ_0. In other words:

$$\begin{cases} yielded \quad region: & |\tau| > \tau_0 \\ unyielded \quad region: & |\tau| \leq \tau_0 \end{cases} \tag{5}$$

Such a conclusion could clarify why conductive regime primes against the convective one when using such complex fluids. The question after that is how to highlight a continuous model for better understanding of such a complex behavior!

Against the discontinuity of this model, various solutions were proposed. Cited the Bercovier and Engelman one in 1980 [43], which based on a regularization introducing within the Bingham viscosity function, such as:

$$\overline{\overline{\tau}} = \left(\mu_p + \frac{\tau_0}{|\dot{\gamma}| + \lambda} \right) \overline{\overline{\dot{\gamma}}} \qquad |\tau| > \tau_0 \tag{6}$$

where λ is a regularization parameter which must be very small compared to $|\dot{\gamma}|$ through the domain. Typically, its value is about $10^{-3}s^{-1}$.

A few years latter, Tanner and Milthorpe [44] regularized the ideal Bingham model with the *bi-viscosity* one, based on two finite viscosity slopes (*i.e. two Newtonian models*) such as:

$$\begin{cases} \mu_{Newtonian} & |\dot{\gamma}| < \dot{\gamma}_c \\ \mu_p \ll \mu_{Newtonian} & |\dot{\gamma}| > \dot{\gamma}_c \end{cases} \tag{7}$$

where $\dot{\gamma}_c$ is the shear rate value of the yield stress τ_0.

Reaching 1984, mathematics was more involved to solve such a complex discontinuity problem. Then, the *Augmented Lagrangian method* ALM was proposed by Glowinski [45] based on the variational inequalities theory. As this complex solution didnt supply convincing results for practical viscoplastic flow applications that time, numerical computations for Bingham problems were dominated once again using a regularization approach. Subsequently, a new solution was proposed by Papanastasiou [46].

In 1987, Papanastasiou proposed an exponential regularization for the Bingham equation (i.e., equation (2)) by introducing the time-dimension parameter m which controls the exponential growth of the stress. Called as *Bingham-Papanastasiou model*, the latter (which judged available for both regions) is presented as follows:

$$\overline{\overline{\tau}} = \left(\mu_p + \frac{\tau_0}{|\dot{\gamma}|} \left[1 - \exp\left(-m \, |\dot{\gamma}| \right) \right] \right) \overline{\overline{\dot{\gamma}}} \tag{8}$$

With a simple shear flow (i.e., 1-D flow), the Bingham-Papanastasiou model is written as follows:

$$\tau = \tau_0 \left[1 - \exp\left(-m\dot{\gamma} \right) \right] + \mu_p \dot{\gamma} \tag{9}$$

Quoted from Mitsoulis and Tsamopoulos [18], and as shown in (Figure 1), the exponent m controls the stress growth, such that below the yield stress τ_0 a finite stress is allowed for vanishingly small shear rates while the stress grows almost linearly with the shear rate and the viscosity μ_p beyond the yield stress. In the limit of $m = 0$, the Newtonian liquid is found to be recovered, when the limit of $m \to \infty$ is fully equivalent to the ideal Bingham model.

On the full tensorial form, the one-dimensional constitutive equation noted above (i.e., equation (9)) can be written as a purely viscous generalized Newtonian fluid, such as:

$$\overline{\overline{\tau}} = \eta \overline{\overline{\dot{\gamma}}} \tag{10}$$

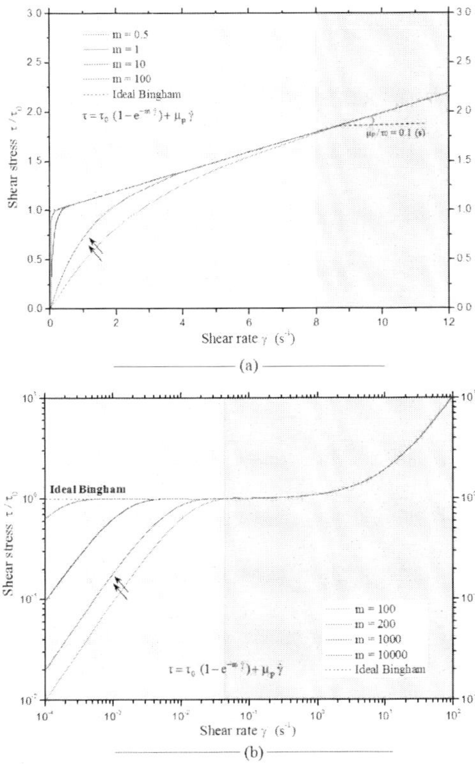

Figure 1. Shear stress vs shear rate for modified Bingham fluids according to the exponential model proposed by Papanastasiou [46] for different values of the regularization parameter m. (a) linear plot, (b) log-log plot.

where η is the apparent viscosity which given, for the Bingham case, as follows:

$$\eta = \mu_p + \frac{\tau_0}{|\dot{\gamma}|}\left[1 - \exp\left(-m\,|\dot{\gamma}|\right)\right] \tag{11}$$

As the viscoplastic character of the flow can be assessed by a dimensionless yield stress [46], which called the Bingham number [16] or Oldroyd number [47] (see equation (12)), it was possible to derive a dimensionless form of the apparent viscosity as well as the Papanastasiou model.

$$\begin{cases} {\tau_0}^* = \frac{\tau_0 H}{\mu_p V_N} \\ H: Characteristic\ length\ (half\ the\ channel\ width\ or\ radius R) \\ V_N: Characteristic\ speed\ taken\ as\ the\ average\ velocity\ of\ a\ corresponding \\ \ Newtonian\ liquid\ with\ viscosity\ at\ the\ same\ pressure\ gradient \end{cases} \tag{12a}$$

$$\begin{cases} Bn = Od = \frac{\tau_0 D}{\mu_p V_m} \\ D: The\ diameter\ or\ channel\ width\ (2H) \\ V_m: The\ average\ velocity\ of\ the\ viscoplastic\ fluid \end{cases} \tag{12b}$$

Scaling the lengths by H, the velocities by V, and the rates-of-strain by V/H; the dimensionless stress growth exponent could be defined as follows:

$$M = \frac{mV}{H} \tag{13}$$

Subsequently, the dimensionless apparent viscosity η^* is written as [48] :

$$\eta^* = 1 + \frac{Bn}{|\dot{\gamma}|}\left[1 - \exp\left(-M\left|\dot{\gamma}\right|\right)\right] \tag{14}$$

what lead us to conclude that the dimensionless criterion to find the yielded and the unyielded regions is the Bingham value (*i.e.*, $|\tau| = Bn$)).

3. Nature of the Porous Matrix

Through this part, both the nature and the physical characteristics of the porous matrix are briefly displayed. The latter, which will be adopted next, consists of the Silicon Carbide Ceramic (*noted as SiC*). Initially observed in 1824 by Jns Jacob Berzelius [49] in a synthetic diamond experience with a parasite reaction between carbon and silice at high temperatures, this porous metal was very used since the industry of 1890, due to its precious properties such as its great thermal conductivity and its powerful resistivity to thermal shocks and chemical oxidation, not to mention its low thermal expansion, to name but a few. As the latter is barely found in its natural form, due to its extreme temperature condition to be created ($\backsim$ 1400 – 1800 °C), a laboratory approach has been adopted to produce such a metal. Subsequently, synthetic methods are mainly developed over the years [51,52] (*See for instance Figure 2*).

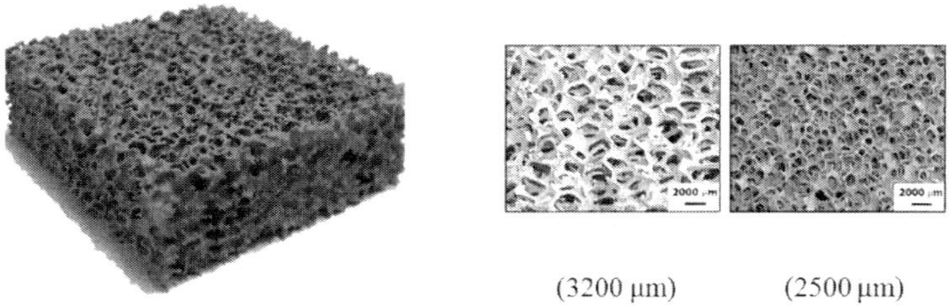

(3200 µm) (2500 µm)

(a) (b)

Figure 2. Silicon carbide foam (a) with various cell sizes (b) [50].

Depending on the synthetic temperature and the reaction time as well, the Silicon Carbide Ceramic (*set by a pilote scale*) could be obtained with various specific surfaces, as between 10 and $40m^2/g$. The table below, Table 1, displays some typical average specific surfaces which computed on spherical, extruded and foam shapes of SiC [50].

Similarly, the conposititon and the granulometry of the used mixture powder can control the porosity and the mean pores diameter of the synthetic SiC. In industrial applications, like filters and catalysts, the porosity could be ranged between 37% and 45% with a mean pore diameter of about 9 to $15\mu m$ [52].

Table 1. Average specific surface of various formes of SiC [50]

SiC shape	Average specific surface $[m^2/g]$
Spheric	10-22
Extrudes	15-35
Foam	10-20

Regarding the permeability, which presents the key factor for most industrial applications, a large range of the latter could be denoted. Summarizing what we found, the industrial permeability of the SiC is denoted between 0.4810^{-9} and 0.9510^{-13} m^2 what gives a Darcy value of about 10^{-3} to 10^{-7}, respectively [52–54].

4. Problem Statement and Mathematical Formulation

4.1. Problem Statement

Using the above information, the developed problem through this chapter consists of a SiC porous medium, which bounded by a cold (and less concentric) outer circular cylinder and a hot (and concentric) inner one. To make a general illustration of such a problem, cylinders are coaxially mounted (*see for instance Figure 3*). Regarding the working fluid that we exploit within the porous matrix, the latter is assumed to be a viscoplastic and obeying the Bingham rheological model [18, 55, 56]. Its thermo-physical properties are presumed to be constant [57, 58] except the density variation within the buoyancy term $\rho_{(T,C)}$ which depends linearly to the local temperature and concentration, as displayed through equation (15) [59]; not to mention the viscosity η which is a function of the yield stress, see for instance equation (11) [57].

$$\rho_{(\mathrm{T,C})}=\rho_0\left[1-\beta_{\mathrm{T}}\left(\mathrm{T}-\mathrm{T}_0\right)-\beta_{\mathrm{C}}\left(\mathrm{C}-\mathrm{C}_0\right)\right] \tag{15}$$

β_T and β_C are the thermal and the concentration expansion coefficients, which can be defined as:

$$\beta_{\mathrm{T}}=-\frac{1}{\rho_0}\left[\frac{\partial\rho}{\partial\mathrm{T}}\right]_{\mathrm{p,C}};\ \beta_{C}=-\frac{1}{\rho_0}\left[\frac{\partial\rho}{\partial\mathrm{C}}\right]_{\mathrm{p,T}} \tag{16}$$

At a general view, the treated solid matrix is assumed to be rigid, isotropic, homogeneous and in thermal equilibrium with the complex fluid [60]. Both, Soret and Dufour impacts on the double diffusion phenomenon are assumed to be insignified [60]. As well, the mass diffusion within the solid matrix is neglected. Regarding the present problem, the permeability of the porous medium is kept uniform. The equivalent Darcy value is about 10^{-3} when the porosity is about 40%.

4.2. Mathematical Formulation

As a numerical investigation, such a physical problem requests a simplified mathematical formulation. Subsequently, adopting the 2D Brinkman extension of the classical Darcy equation brings to light general conservation equations, which can describe the transport phenomenon of the Bingham fluid within the porous annulus as follows:

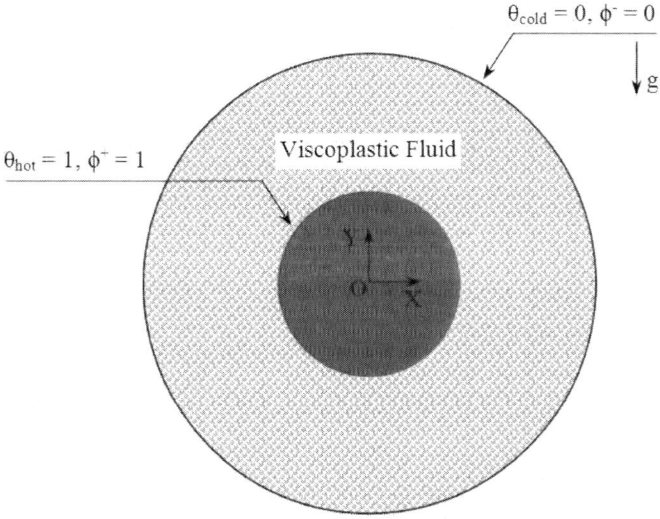

Figure 3. Simulation domain with its boundary conditions.

$$\vec{\nabla}\overrightarrow{V} = 0 \quad (17a)$$

$$\rho_0 \frac{\partial \overrightarrow{V}}{\partial t} + \frac{\rho_0}{\varepsilon^2}\left(\overrightarrow{V}.\vec{\nabla}\right)\overrightarrow{V} = \rho\overrightarrow{g} - \vec{\nabla}p - \frac{\eta}{K}\overrightarrow{V} + \vec{\nabla}.\overline{\overline{\tau}} \quad (17b)$$

$$\frac{\partial T}{\partial t} + \overrightarrow{V}.\vec{\nabla}T = \left(\frac{k}{\rho_0 C_p}\right)^{eq} \nabla^2 T \quad (17c)$$

$$\frac{\partial C}{\partial t} + \overrightarrow{V}.\vec{\nabla}C = D^{eq}\nabla^2 C \quad (17d)$$

where ϵ, C_p^{eq}, k^{eq} and D^{eq} displays the porosity, the equivalent specific heat, thermal conductivity and the mass diffusivity, respectively.

One important to note, that the plastic viscosity μ_p as well as the yield stress τ_0 are assumed to be independent to face the temperature, due to the experimental evidence that the yield stress is approximately independent to the temperature variation and the plastic viscosity is a barely decreasing function of the latter (i.e., *the temperature*) for a well-known yield stress system in the temperature range of 0 to $90°C$ [61].

5. Numerical Modelisation and Code Validity

5.1. Numerical Procedure

The *Cartesian Cut-Cell* approach, also called Cartesian Grid method [62, 63], is adopted to handle the limits of our circular cylinders (*see Figure 4*). As such, the control volumes which found within the inner cylinder are assumed to be inactive following the inequality equation (18). Simply said, neither the movement of our convective fluid nor its thermosolutal transfer are denoted within the inner source ($i.e., U = V = 0, T = T_h, C = C^+$). As well, the same condition is imposed to the outside part of the outer cylinder as shown Equations (18) ($i.e., U = V = 0, T = T_c, C = C^-$)

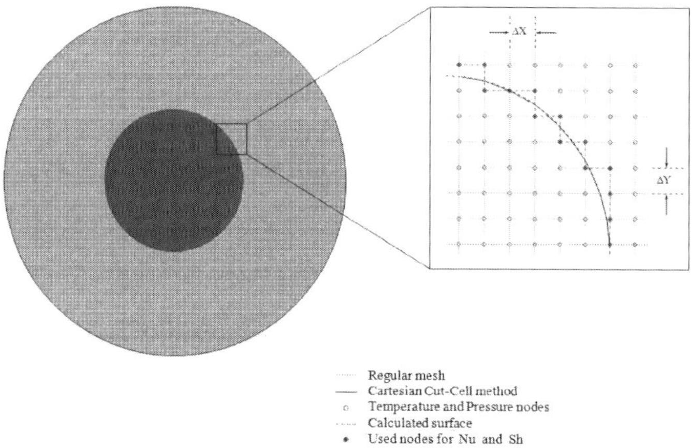

Figure 4. Grid arrangement and Cartesian Cut-Cell method.

$$(\mathrm{x} - 0.5)^2 + (\mathrm{y} - 0.5)^2 \le \left(\frac{D_i}{2}\right)^2, \ U = V = 0 \ , \ T = T_h \ , C = C^+ \tag{18a}$$

$$(\mathrm{x} - 0.5)^2 + (\mathrm{y} - 0.5)^2 \ge \left(\frac{D_o}{2}\right)^2, \ U = V = 0 \ , \ T = T_c \ , C = C^- \tag{18b}$$

Once the physical domain is defined, the governing conservation equations are discretized in space using the finite volume approach, when the convective and the diffusive terms were treated using a Power-Law scheme. The resulting algebraic equations side by side with the associated boundary conditions are solved, consequently, using the line by line method [64].

As the momentum equation is expressed in terms of the various primitive variables (U, V and P), the iterative procedure includes a pressure correction calculation method, namely SIMPLER [64], to solve the pressure-velocity coupling. Compared to other pressure-velocity coupling approaches, such as SIMPLE or SIMPLEC, the adopted approach has proven to be fastest (*about* 30 *to* 50% *fewer iterations*) [64].

As a fundamental numerical condition, the convergence criterion for the computed variables such as the temperature, the concentration, the pressure and the velocity as well, is defined as follows:

$$\frac{\sum_{j=1}^{m}\sum_{i=1}^{n}\left|\xi_{i,j}^{A+1}-\xi_{i,j}^{A}\right|}{\sum_{j=1}^{m}\sum_{i=1}^{n}\left|\xi_{i,j}^{A+1}\right|}\leq 10^{-5} \tag{19}$$

where both m and n are the values of the grid points in X and Y directions, respectively, ξ is any of the computed field variables, and A is the iterations number.

Regarding the small cells that could be found arbitrarily cut; reason to the Cartesian Cut-Cell method Colella et al. [63] proposed to compute the fluid flow in the first step without taking these interfaces into account. Then, the loss interfaces can be calculated and distributed to the neighboring cells, based on the fluid mass which presents in each.

5.2. Brief Code Validation

For years, the performance of the home-developed code against the heat conduction phenomenon and the mass diffusion one was accomplished within various areas and boundary conditions, by comparing our predictions with other numerical results and experimental data, already available in the literature, and by verifying the grid independence of the present results [65–75]. Subsequently, only a brief validation discussion is displayed through this chapter.

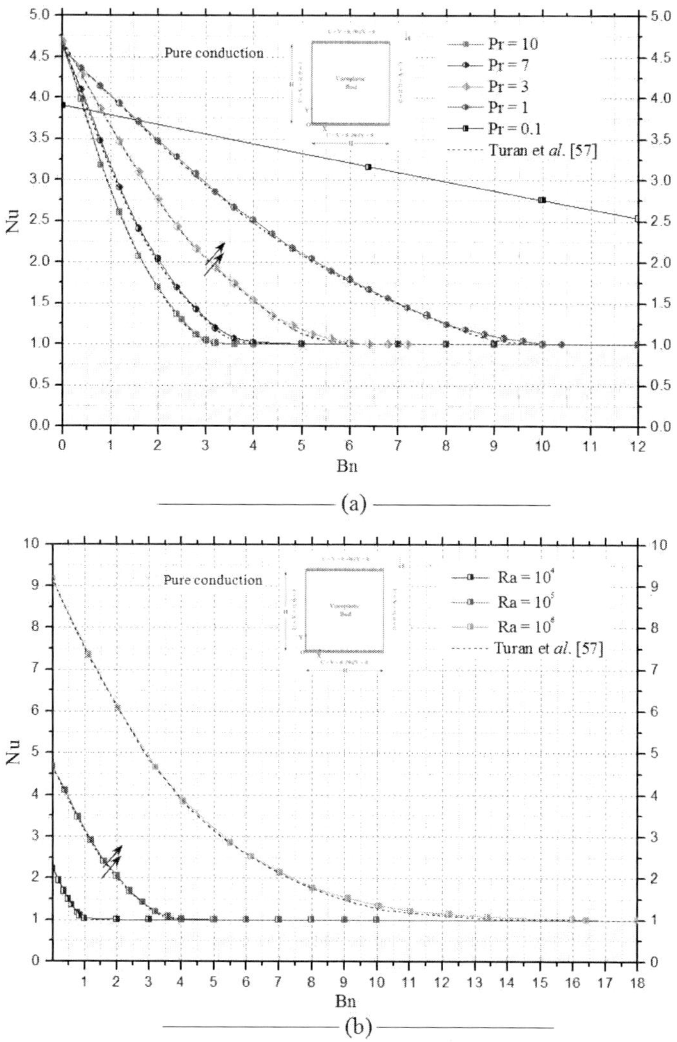

Figure 5. Viscoplastic fluid validation in terms of the mean heat transfer: (a) mean Nusselt number as a function of Prandtl number, $Ra = 10^5$. (b) mean Nusselt number as a function of Rayleigh number, $Pr = 7$.

Starting with the first one which consists the previous computations of Turan al. [57,58], who deals with natural convection of a Bingham fluid within square enclosures. By taking into account the same hypotheses, Figure 5(a,b) presents the computed transfer rate of various Bingham values relative to Prandtl and Rayleigh numbers, respectively. As we can see, the present results and those of Turan et al. [57,58] are in excellent agreement with a maximum discrepancy of about 1%.

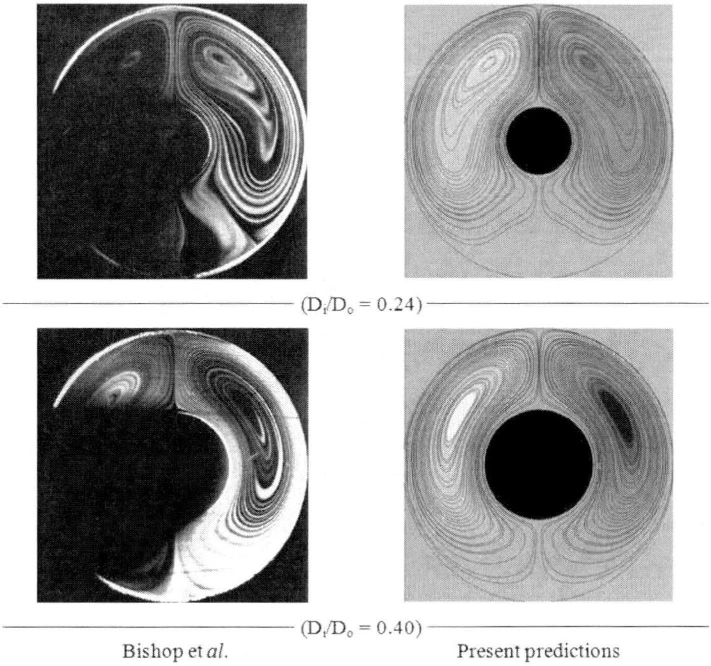

Figure 6. Experimental streamlines validation, Bishop et al. [76].

To ascertain its validity with a cylindrical annulus shape, experimental predictions of Bishop et al. [76] and Grigull et al. [77] have been taken into consideration. Figures 6 and 7 displayed the comparison between these data and our predictions in terms of streamlines and temperature contours, respectively. Once again, the results exhibit a great qualitative concordance with the experimental ones what supports the exploitation of the code to investigate our phenomenon in such a configuration.

Considering the appropriate grid for the next simulation part, a grid independence investigation was conducted for double diffusive convection of a Bignham

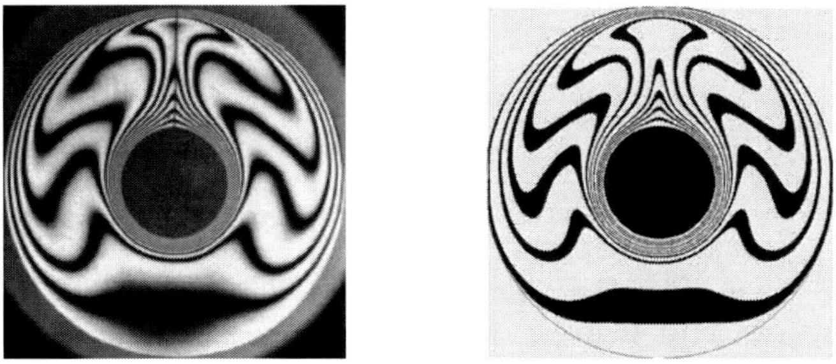

Grigull et *al.* Present predictions

Figure 7. Experimental isotherms validation, Grigull et al. [77].

fluid within the porous annulus. As such, several mesh distributions were tested for a Bingham value of about 4 as long as the central V-velocity profile is evaluated (*see for instance* Figure 8). Subsequently, reaching a uniform mesh grid of about 201^2 highlights an independent grid solution. Nevertheless, a fine structured mesh of 401^2 is judged to be well used, both for the sake of the physical phenomenon safety and to avoid round-off error for all other calculations in our investigation.

6. Results and Discussion

6.1. About the Order of Magnitude Analysis

Over the years, thermosolutal convection within a porous annulus has been treated in our papers for a large range of relevant parameters [14, 15, 65, 66]. Cited the fluid character,*which can be presented by the Bingham number* $(0 \leq Bn \leq 50)$, the boundary layers ratio, *known generally as Lewis number* $(1 \leq Le \leq 300)$, the porous buoyancy, *which known as the Rayleigh-Darcy combination* $((100 \leq Ra_T Da \leq 2000)$, as well as the geometry aspect ratio $(20\% \leq Di/Do \leq 96\%)$ and the thermosolutal buoyancy one, *which is a pure relation between the thermosolutal gradients and its expansion coefficients* $(-50 \leq (\beta_C \Delta C)/(\beta_T \Delta T) \leq 50)$, to name but a few.

Against such a great number of the pertinent parameters, an order of magnitude analysis proved to be very complicated when it comes to define the best

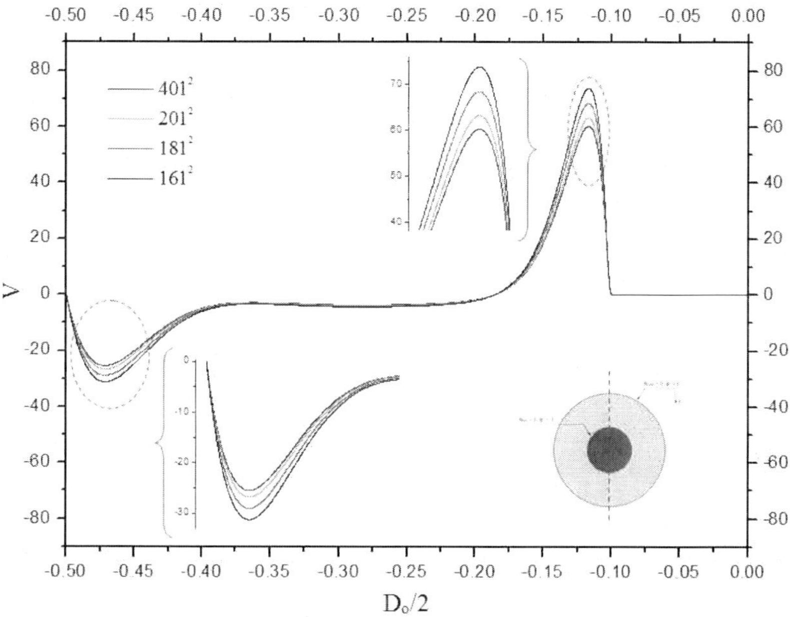

Figure 8. V-velocity profile along the mid-section of the porous annulus for various uniform grids. $Bn = 4$, $Di/Do = 0.20$, $Le = 10$, $N = 10$, $Pr = 10$, $Ra^* = 100$.

conditions for a straight prediction of such a double diffusive phenomena. Reason to this, it was very useful to structure such a scale analysis as denoted by Trevisan and Bejan [78], Bghein et al. [79], Bennacer et al. [29], Gobin et al. [80], to namebut a few.

One way to do is to recognize the two boundaries between which, our system could evolve [78, 79]. Then, we denote a fluid flow which is mainly caused by a heat transfer, as the buoyancy impact is primarily due to the temperature changes (see for instance equation (20)).

Conversely, the fluid flow could be caused by a solutal transfer. As a result, the buoyancy impact is dominated by density changes which is related to the concentration variations equation (20):

$$|\beta_T \Delta T| \gg |\beta_C \Delta C| \tag{20a}$$
$$|\beta_C \Delta C| \gg |\beta_T \Delta T| \tag{20b}$$

Consequently, the transition from the flow of case (01) to that of case (02) is dominated by the buoyancy ratio $(\beta_C \Delta C)/(\beta_T \Delta T)$.

The main question was about the nature of that double diffusive buoyancy within such a transition phase. In their work, Bennacer and Gobin [29, 80] proved that, for a Lewis value greater than unity $(Le > 1)$, this new area could be divided into two sections: as a solutal transfer which governed by a thermal buoyancy (1st case);and a thermal transfer which controlled by a solutal buoyancy (2nd case).

Their numerical predictions, for a Newtonian fluid case, implies that when we leave the dominant thermal regime to a compositional thermal-solutal dominated one, following the increase of the buoyancy ratio, the solutal transfer rate goes to a new regime in which, its value will not only a function of $|N|$ but to the ratio $N/Le^{1/3}$:

$$\frac{(Sh)_C}{(Sh)_T} = \frac{(\delta_C)_T}{(\delta)_C} \sim \left(\frac{N}{Le^{1/3}}\right)^{1/4} > 1 \tag{21}$$

As for the latter, the solutal-thermal composition regime is proved to be defined by the ratio N/Le. Such crucial results were the key to go far and wondering about the domination buoyancies in such a transition section within a pure fluid or a porous medium.

As for the space nature, using a viscoplastic fluid has no significant effect of the limits of the transition section, i.e., the ratio N/Le, which can be defined as $N/Le^{1/2}$. To make clear such a conclusion, Figure 9 displays the mean solutal transfer as a function of the ratio $N/Le^{1/2}$ and that, for a Bingham fluid case that filled a differentially heated and concentred porous square.

6.2. About the Conduction and the Thermosolutal Stability-Limit

As far as the order of magnitude analysis can presume the impact of each boundary layer on the domination of the thermal or the solutal transfer, it is still very hard to make clear the combined diffusions impact on the fluid flow. The question, then, is how to expound such a point, especially when we talk about a pure conduction transfer or a mass diffusive one!

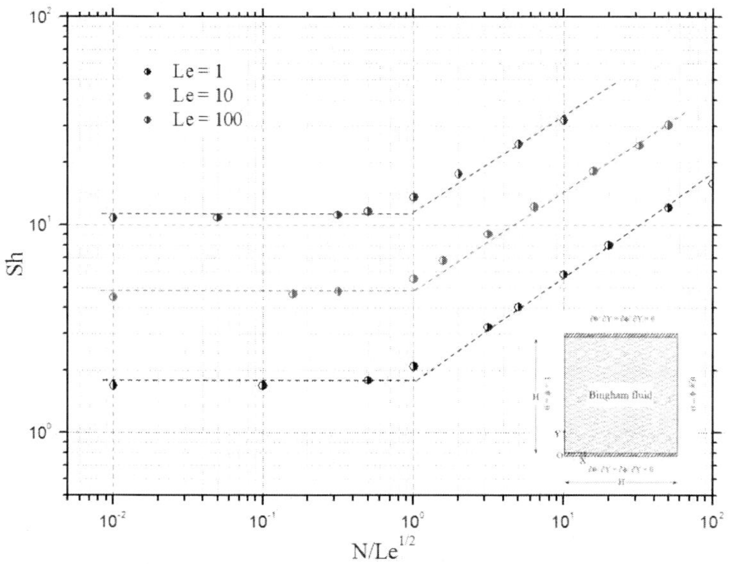

Figure 9. Impact of the ratio $N/Le^{0.5}$ on the mass transfer rate. $Bn = 5$, $Pr = 10$, $Ra = 10^5$.

6.2.1. Impact of the Boundary Layers Ratio

As displayed in the above sections, the SiC was chosen to be our porous matrix for which, we denote the Darcy and the porosity values as 10^{-3} and 40%, respectively. During our predictions, the computed ratio between the thermal diffusivity forces and the viscous diffusivity ones is kept constant and equal to 10 (simply said, $\nu/\alpha = Pr = 10$). Then, an aspect ration of about 20% is used, which is good enough to avoid the oscillatory phenomena within the porous annulus for the adopted range of the modified Rayleigh number [81, 82].

As a first description, the Algerian crude oil is chosen to be the convective fluid inside the porous annulus. The corresponding Bingham value at an ambient temperature ($T \approx$ 20 to $28°C$) is found to be in between 0.85 and 92, depending to the water quantity which can be presented within [83–85]. Subsequently, a Bingham value of about 4.96 is fixed to ensure the mono-phase case [83–85]. At this value, Figure 10 displays a series of hydrodynamic fields side by side with thermal and solutal ones according to various Lewis values ($Le \sim \delta_T/\delta_C$) and that, for a buoyancy ratio N and a porous buoyancy $Ra_T \times Da$ of about 10 and 100, respectively. Consequent to the symmetrical vertical plane ($-0.5 \leq Y_{X=0} \leq +0.5$), only the half part of these profiles is presented.

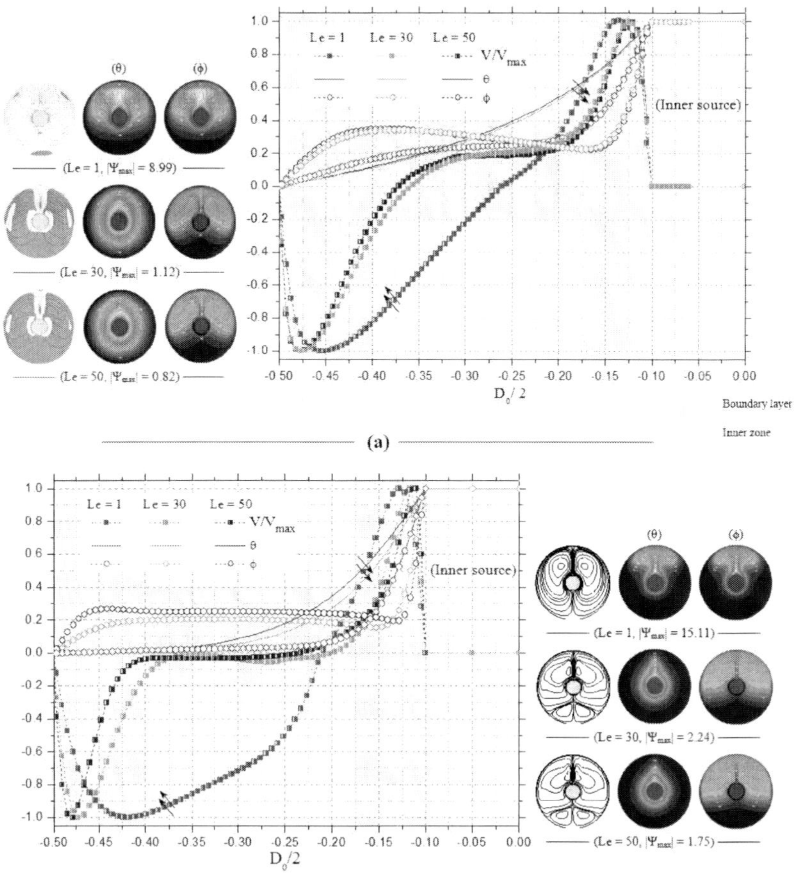

Figure 10. V/V_{max}, θ and ϕ profiles along the horizontal mid-plane of the porous annulus for various Lewis values. (a) $Bn = 4.96$, (b) $Bn = 0$, $D_i/D_o = 0.20$, $N = 10$, $Ra^* = 100$.

The primary examination of the velocity profile V/V_{max} at various Lewis values proved the existence of a clockwise and an anti-clockwise circulating cells within the annulus, as displayed the corresponding streamlines. Except the case of $Le = 1$, the velocity of the Bingham fluid near the active cylinders is mainly developed by a pure thermosolutal buoyancy, which caused by a combined impact of both the temperature and the concentration variations. Such a conclusion was very difficult to make on the basis of the corresponding isotherms and iso-concentrations, especially when the plug-flow character of the complex fluid (i.e., *the gray area*) takes a full control within the annulus for an increasing Lewis value.

The knowledge obtained by investigating the limit case of the Bingham-plastic fluid, i.e., *the Newtonian fluid*, brings to light a new stability-limit of such a double diffusive convection. In fact, and even with the absence of the plug flow character, the dominant conduction mode is always present within the annulus as long as the Lewis value gets important. Nevertheless, only the pure solutal buoyancy can control the flow near the outer cylinder, as the temperature variation is found to be insignificant through that area. Hence, we could talk about a transition phase from a solutal-thermosolutal domination behavior to a pure thermosolutal one by increasing Bingham number. At this part, we start wondering about the impact of the buoyancy ratio on such results, as will be discussed in the section below.

6.2.2. Impact of the Buoyancy Ratio

Going far with our predictions, and when the Lewis value is set at 10, Figure 11 displays the impact of the buoyancy ratio on the Bingham flow field, side by side with its thermal and solutal behaviors. Once again, two symmetrical counter-rotating cells are developed within the porous annulus. Its absolute value of stream-function proves to be an increasing function of the buoyancy ratio, due to the fluid motion within the porous matrix which gets powerful. Unlike the case before, the isotherm plots are more pronounced with increasing buoyancy ratios, indicating the transition from a conductive mode (*as for* $N = 5$) to a primary-convective one (*as for* $N = 30$).

As to the concentration field, the thickness of the mass boundary layer near the active cylinders turns to be a decreasing function of the enhanced buoyancy ratio. Such a behavior reflects a dominant diffusion mechanism within the porous annulus. Consequently, the rigid structure of the Bingham fluid that ap-

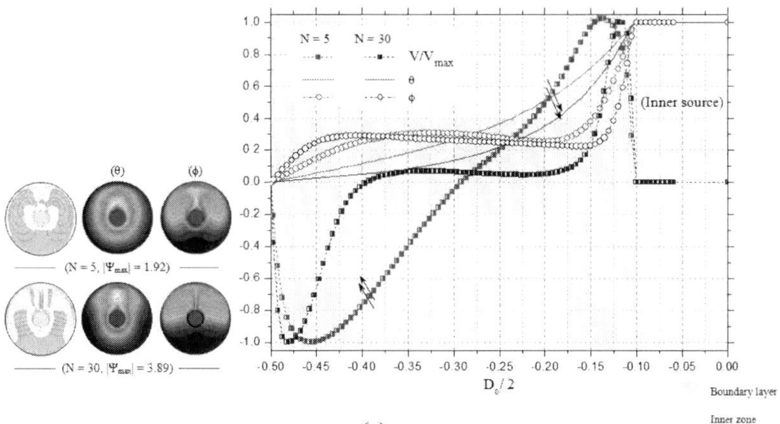

(a)

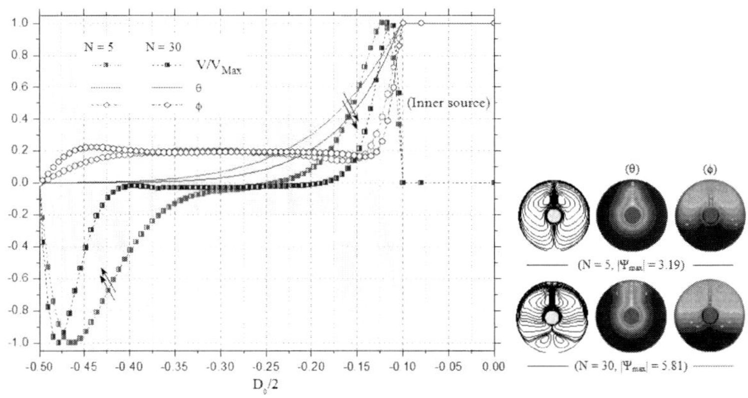

(b)

Figure 11. V/V_{max}, θ and ϕ profiles along the horizontal mid-plane of the porous annulus for various buoyancy ratio values. (a) $Bn = 4.96$, (b) $Bn = 0$, $D_i/D_o = 0.20$, $Le = 10$, $Ra^* = 100$.

pears for a low buoyancy value could be easily broken by improving the values of the buoyancy ratio.

Complement to these conclusions, the hydrodynamic boundary layers near the active cylinders turn to be developed by the pure thermosolutal buoyancy for moderate buoyancy ratios. Unlike, and as far as the buoyancy ratio gets important (*as for* $|N| = 30$), only the solutal gradient governs the flow near the outer cylinder since the temperature variation is insignificant within that area.

Regarding the Newtonian case, the same stability limit is denoted with increasing values of the buoyancy ratio. Once again, and even with the primary heat transition which denoted from moderate buoyancies to the important ones, only the pure solutal buoyancy can control the flow near the outer cylinder. Accordingly, we can keep talking about a transition phase from a solutal-thermosolutal dominant behavior to a pure thermosolutal one by increasing Bingham number, but only with a specific range of the buoyancy ratio this time.

6.2.3. The Transition Phase Against the Annulus Aspect Ratio

Complement to what was previously presented, and as long as the annulus aspect ratio is ranged between 20% and 96% our predictions are regrouped in Figure 12 depending on the fluid character, i.e., *Bingham values*. Then, the ratio of the boundary layers is set fixed at 10, when the buoyancy one is equal to 5, what keeps the critical ratio $N/Le^{0.5}$ barely greater than unity which is good enough to avoid the pure dominant conduction (*see for instance Figure 9)*.

Subsequently, a further increase in the Bingham value for a given aspect ratio reveals a critical Bingham value for which, the transition from a solutal-thermosolutal dominant phenomenon to a pure thermosolutal one is easily detected.

Going far with its value, the increase in the annulus aspect ratio to 80% causes a significant modification on such a transition phase, as the pure thermosolutal domination phenomena takes the main control by increasing values of the ratio D_i/D_o. Such a complex link between the fluid nature, the aspect ratio of the annulus, and the transition phase limits might prove particularly useful in verifying any instability analysis which might be put forth in the future.

6.3. About the Heat Conduction and the Oscillatory Flow-Limit

By taking our investigation to another level, the oscillatory flow within annulus is concerned though this part of the chapter. As reported in several experiments and numerical works [76, 86–88] and based on the thermal gradient effect, it has been established that the basic fluid flow at low Rayleigh values consists of two symmetrical cells within the confined space, relative to the vertical plane that containing the cylinder axis [89]. As long as the Rayleigh value gets important

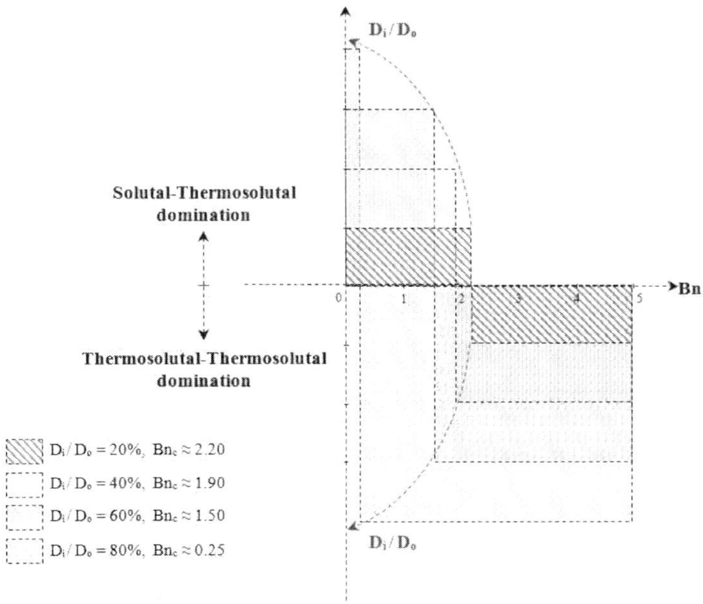

Figure 12. The transition limit against the annulus aspect raio for various Bingham numbers.

(i.e., *the thermal buoyancy*), such a flow becomes unstable, what leads to a very complicated phenomenon called *Oscillation*.

Against such a flow behavior, various experimental and numerical studies have been developed over the years. The main complexity was to find the perfect expression that can relate the impact of both the thermal buoyancy and the annulus aspect ratio to the oscillation limit. Contradictions in the description of these flows prompted a study directed at obtaining more detailed information regarding the onset and the main characteristics of this oscillatory flow. Our brief investigation is undertaken to extend the Rayleigh number range with the critical aspect ratio of the annulus, in conjunction with the extension of this range, to obtain precise information about the oscillation characteristics.

To better understand this complex phenomenon, a three-dimensional prediction of a Newtonian fluid will be reported through this part. Both the Lewis value and the buoyancy ratio one are set at 1 and 0, respectively (i.e., *an identical thermal-solutal diffusion with an absent solutal buoyancy*).

6.3.1. Heat Conduction Limit at Fluid Annulus

As far as the Darcy value is greater than unity ($Da \geq 1$), Figure 13 displays a series of streamlines and isotherm plot according to various thermal buoyancies, such as $Ra = 10^3$ to 10^6 and that, for an aspect ratio of about 24% which is good enough to generate an oscillatory phenomenon [76, 82].

Subsequently, two symmetrical counter-rotating celle are formed within the annulus. The fluid flows up along the inner-cylinder surface and down along the outer-cylinder one at a relatively high speed with increasing Rayleigh value (i.e., *the thermal buoyancy*).

The represented isotherms for moderate thermal buoyancies, (*i.e.,* $Ra = 10^3$ and $Ra = 10^4$), turn to be barely changed, demonstrating a dominated conduction regime, unlike the iso-plots of a high Rayleigh value, such as $Ra = 10^5$ and $Ra = 10^6$, where the latter are more distorted. Simply said, a thermal boundary layer is shaped all along the active surfaces and becomes thinner with high Rayleigh values. In addition to this behavior, a thermal plume appears in the upper portion of the inner cylinder (*see for instance the case of* $Ra = 10^6$) what proved the domination of the convection mechanism within the confined space.

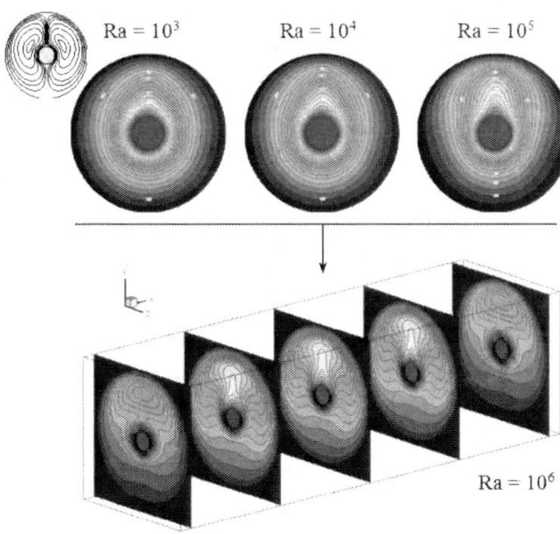

Figure 13. 3D Visualization of streamlines and isotherm plots for various Rayleigh values. $Pr = 6.2$, $D_i/D_o = 0.24$.

In fact, what we presented through the previous figure (Figure 13) is only a fixed solution of the conduction history within the simulation time in which, the oscillatory phenomenon hasnt started yet. Going far with the unsteady solution, Figure 14 comes out to make clear the velocity evolution with thermal buoyancy. The latter, which computed at a specific positions ($X = 0.5$, $Y = 0.3$ and $Z = 0.$) and presented with various Rayleigh values, proved that the flow could be developed at a stable regime (i.e., *steady solution*) under a critical value of thermal buoyancy. Such a limit was defined with a Rayleigh value of about $9.9\,10^5$.

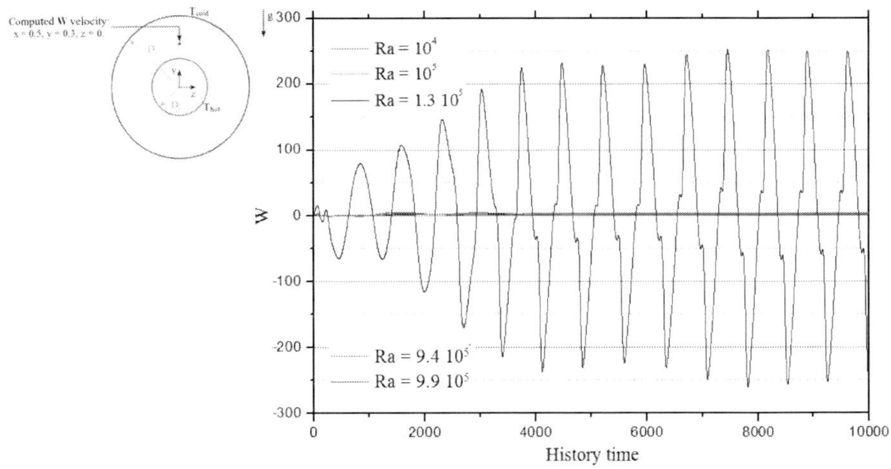

Figure 14. Thermal buoyancy and the oscillation flow behavior. $D_i/D_o = 0.24$, $Pr = 6.2$.

6.3.2. Oscillation Stability Limit

Yes of course, the critical Rayleigh value is related to a fixed aspect ratio, what makes us wonder about the impact of the latter (i.e., the aspect ratio) as long as we operate various buoyancy values. To make this point clear enough, the evolution of the velocity profile with various aspect ratios are displayed through Figure 15 for a Rayleigh value of about 10^6. As such, using an aspect ratio greater than 24% makes the oscillation phenomenon very intermittent. At a

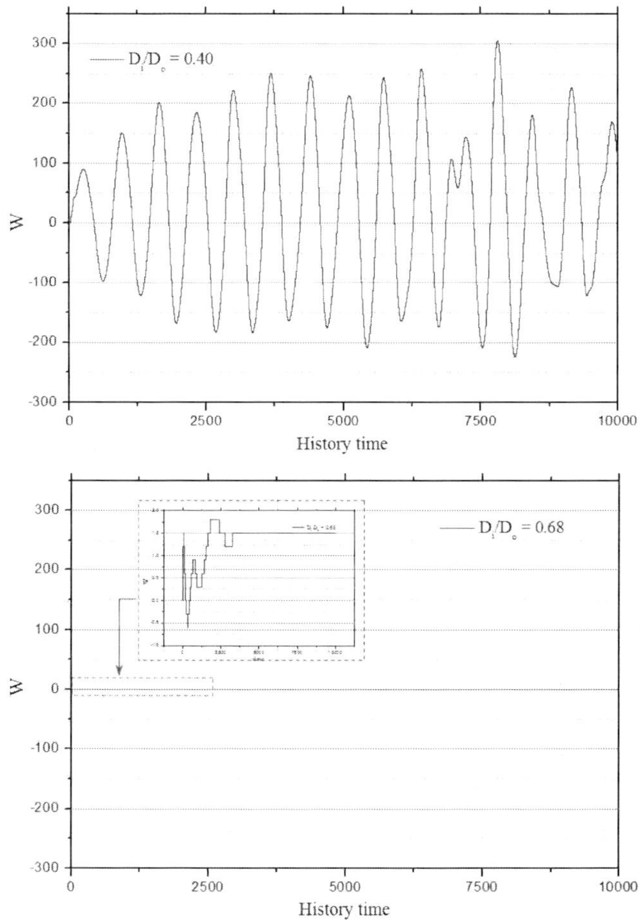

Figure 15. Velocity profiles as a function of the oscillation history time for various aspect ratios, $Ra = 10^6$, $Pr = 6.2$.

critical value of about 64%, such a complex phenomenon disappears to let the stable regime rise again within the confined space. Hence, we could talk about a critical steady state sotution of our transfer phenomenon.

Conclusion

Through this chapter, the conduction limit of a double diffusive convection of a Bingham fluid was numerically treated within a cylindrical SiC annulus. Following the impact of some relevant parameters, as the character of the viscoplastic material, the boundary layers ratio, as well as the thermosolutal buoyancy one, such a critical limit was widely inspected.

Subsequently, such a thermosolutal contribution on the flow field of the viscoplactic fluid turned out to be very complicated, compared to the pure convection or solutal diffusion within the porous space. This complexity grows as long as the impact of the Lewis and buoyancy ratio one are combined at low porous buoyancies. Consequently, follow-up the development of the boundary layers near the active walls was highly required in order to enlighten such a complex contribution.

At the light of our predictions, a transition period from a solutal-thermosolutal phenomenon to a pure thermosolutal one has been defined at a specific buoyancy range. Such a complex zone, which is a function of both the fluid nature and the geometry aspect ratio, could be considered as a complement to some previous predictions of Newtonian fluids. The latter might prove particularly useful in verifying any instability analysis which might be put forth in the future.

Going far in the predictions, the oscillatory character within such an industrial configuration was inspected as well. By taking into account a large range of the thermal buoyancy and the geometry aspect ratio, such a noising flow phenomenon could be only controlled by using a conjugate effect of these relevant parameters. Consequently, critical conditions have been set to allow the steady state to rise easily within the annulus.

References

[1] Nield D. A. and Bejan A., *Convection in porous media.* Springer. 2006.

[2] Mamou M., Vasseur P., Bilgen E., Multiple solutions for double-diffusive convection in a vertical porous enclosure, *Int. J. Heat Mass Transfer* **38**, 1787–1798 (1995).

[3] Nithiarasu P., Seetharamu K. N., Sundararajan T., Double-diffusive natural convection in an enclosure filled with fuid-saturated porous medium: a

generalized non-darcy approach, *Num. Heat Transfer, Part A Appl.* **30**, 413–426 (1996).

[4] Ostrach S., Natural convection with combined driving forces, *Physic. Chemical Hydrodynamics* **01**, 233–247 (1980).

[5] Kamotani Y., Wang L. W., Ostrach S., Jiang H. D., Experimental study of natural convection in shallow enclosures with horizontal temperature and concentration gradients, *Int. J. Heat Mass Transfer* **28**, 165–173 (1985).

[6] Lee J., Hyun M. T., and Kim K. W., Natural convection in confined fluids with combined horizontal temperature and concentration gradients, *Int. J. Heat Mass Transfer* **31**, 1969–1977 (1988).

[7] Benard C., Gobin D., Thevenin J., Thermosolutal natural convection in a rectangular enclosure: numerical results, *In Heat Transfer Conv. Flows* **107**, 249–254 (1988).

[8] Han H. and Kuehn T. H., Double diffusive natural convection in a vertical rectangular enclosure. numerical study, *Int. J. Heat Mass Transfer* **34**, 461–471 (1991).

[9] Chang J. and Lin T. F., Unsteady thermosolutal opposing convection of liquid-water mixture in a square cavity (II) flow structure and fluctuation analysis, *Int. J. Heat Mass Transfer* **36**, 1333–1345 (1993).

[10] Chen F. and Chen C. F., Double-diffusive fingering convection in a porous medium, *Int. J. Heat Mass Transfer* **36**, 793–807 (1993).

[11] Trevisan O. V. and Bejan A., Mass and heat transfer by high Rayleigh number convection in a porous medium heated from below, *Int. J. Heat Mass Transfer* **30**, 2341–2356 (1987).

[12] Lin T. F., Huang C. C., Chang T. S., Transient binary mixture natural convection in square enclosures, *Int. J. Heat Mass Transfer* **33**, 287–299 (1990).

[13] Bennacer R., *Thermosolutal convection: fluid flow and heat transfer numerical simulations*, Pierre and Marie Curie, Ph.D. Thesis, France, 1993.

[14] Ragui K., Boutra A., Benkahla Y. K., On the validity of a numerical model predicting heat and mass transfer in porous square cavities with a bottom thermal and solute source: case of pollutants spreading and fuel leaks, *Mech. Indust.* **17**, 311 (2016).

[15] Ragui K., Boutra A., Bennacer R., Labsi N., Benkahla Y. K., Correlating heat and mass transfer coefficients for thermosolutal convection within a porous annulus of a circular shape: case of internal pollutants spreading, *J. Heat Mass Transfer* **01**, 01–18 (2018).

[16] Bingham E. C., *An investigation of the laws of plastic flow*, Govt. Print. Off., 1917.

[17] Bingham E. C., *Fluidity and plasticity*, McGraw-Hill, 1922.

[18] Mitsoulis E. and Tsamopoulos J., Numerical simulations of complex yield-stress fluid flows, *Rheologica Acta* **56**, 231–258 (2017).

[19] Pascal H., Rheological behaviour effect of non-newtonian fluids on steady and unsteady flow through a porous medium, *Int. J. Num. Analytical Meth. Geomechanics* **07**, 289–303 (1983).

[20] Amari B., Vasseur P., Bilgen E., [Natural convection of non-newtonian fluids in a horizontal porous layer] naturliche konvektion nicht-newtonscher fluide in einer horizontalen porosen schicht, *Warme-und stoffubertragung* **29**, 185–193 (1994).

[21] Oukhlef A., *Determination de la distribution de tailles de pores d'un milieu poreux* [*Determination of the pore size distribution of a porous medium*], Arts et Mtiers, Ph.D. Thesis, ParisTech, 2011.

[22] Dos Santos D. D., Frey S., Naccache M. F., P.R. de Souza Mendes, Numerical approximations for flow of viscoplastic fluids in a lid-driven cavity, *J. Non-Newt. Fluid Mech.* **166**, 667–679 (2011).

[23] Labsi N., Benkahla Y. K., Boutra A., Study of convective heat transfer in a square cavity filled with a viscoplastic fluid by taking into account viscous dissipation, *Heat Trans. Res.* **44**, (2013).

[24] Boutra A., Benkahla Y. K., Ameziani D. E., Bennacer R., Lattice boltzmann simulation of natural convection in cubical enclosures for the Bingham plastic fluid, *Heat Trans. Res.* **48**, (2017).

[25] Min T., Choi H. G., Yoo J. Y., Choi H., Laminar convective heat transfer of a bingham plastic in a circular pipe (II). Numerical approach hydrodynamically developing flow and simultaneously developing flow, *Int. J. Heat Mass Transfer* **40**, 3689–3701 (1997).

[26] Nascimento U. C. S., Macedo E. N., Quaresma J. N. N., Thermal entry region analysis through the finite integral transform technique in laminar flow of bingham fluids within concentric annular ducts, *Int. J. Heat Mass Transfer* **45**, 923–929 (2002).

[27] Kamotani Y., Wang L. W., Ostrach S., Jiang H. D., Experimental study of natural convection in shallow enclosures with horizontal temperature and concentration gradients, *Int. J. Heat Mass Transfer* **28**, 165–173 (1985).

[28] Trevisan V. O. and Bejan A., Natural convection with combined heat and mass transfer buoyancy effects in a porous medium, *Int. J. Heat Mass Transfer* **28**, 1597–1611 (1985).

[29] Bennacer R. and Gobin D., Cooperating thermosolutal convection in enclosures (I): scale analysis and mass transfer, *Int. J. Heat Mass Transfer* **39**, 2671–2681 (1996).

[30] Bird R. B., Dai G. C., Yarusso B. J., The rheology and flow of viscoplastic materials, *Rev. Chem. Eng.* **01**, 01–70 (1983).

[31] Nield D. A., Bejan A., Bejan N., *Convection in porous media*, Springer, 2006.

[32] Mamou M., Vasseur P., Bilgen E., Multiple solutions for double-diffusive convection in a vertical porous enclosure, *Int. J. Heat Mass Transfer* **38**, 1787–1798 (1995).

[33] Nithiarasu P., Seetharamu K. N., Sundararajan T., Double-diffusive natural convection in an enclosure filled with fluid-saturated porous medium: a generalized non-darcy approach, *Num. Heat Transfer, Part A Appl.* **30**, 413–426 (1996).

[34] Herschel W. H. and Bulkley R., Konsistenzmessungen von GummiBenzol-Loesungen. [Consistency Measurements of Rubber Benzene Solutions] *Kolloid Z, CrossRef Google Scholar* **39**, 291–300 (1926).

[35] Casson N., *Rheology of disperse systems*, Pergamon Press, London, 1959.

[36] Covey G. H. and Stanmore B. R., Use of the parallel-plate plastometer for the characterisation of viscous fluids with a yield stress, *J. Non-Nowt. Fluid Mech.* **08**, 249–260 (1981).

[37] Dzuy N. Q. and Boger D. V., Yield stress measurement for concentrated suspensions, *J. Rheo.* **27**, 321–349 (1983).

[38] Dzuy N. Q. and Boger D. V., Direct yield stress measurement with the vane method, *J. Rheo.* **29**, 335–347 (1985).

[39] Nguyen Q. D. and Boger D. V., Characterization of yield stress fluids with concentric cylinder viscometers, *J. Rheo.* **26**, 508–515 (1987).

[40] Nguyen Q. D. and Boger D. V., Measuring the flow properties of yield stress fluids, *Annual Rev. Fluid Mech.* **24**, 47–88 (1992).

[41] Oldroyd J. G., A rational formulation of the equations of plastic flow for a bingham solid, In: *Math. Proceed. Cambridge Philosophical Society* **43**, Cambridge University Press, 100–105 (1947).

[42] Oldroyd J. G., Two-dimensional plastic flow of a bingham solid: a plastic boundary-layer theory for slow motion., In: *Math. Proceed. Cambridge Philosophical Society* **43**, Cambridge University Press, 383–395 (1947).

[43] Bercovier M. and Engelman M., A finite-element method for incompressible non-newtonian flows, *J. Comp. Physics* **36**, 313–326 (1980).

[44] Tanner R. I. and Milthorpe J. F., Numerical simulation of the flow of fluids with yield stress, In: *Proc. 3rd Int. Conf., Seattle, Pineridge Press, Swansea* **01**, 680–690 (1983).

[45] Glowinski R. and Oden J. T., Numerical methods for nonlinear variational problems, *J. Appl. Mech.* **52**, 739–749 (1985).

[46] Papanastasiou T. C., Flows of materials with yield, *J. Rheo.* **31**, 385–404 (1987).

[47] B.D. de Besses, Magnin A., Jay P., Sphere drag in a viscoplastic fluid, *AIChE. J.* **50**, 2627–2629 (2004).

[48] Huilgol R. R., Variational inequalities in the flows of yield stress fluids including inertia: theory and applications, *Phys. of Fluids.* **14**, 1269–1283 (2002).

[49] Berzelius J., Annalen der Physik [Annals of Physics], *Wiley Edit. Serv.* **169**, 1984.

[50] Leofanti G., Padovan M., Tozzola G. and Venturelli B., Surface area and pore texture of catalysis, *Catalysis Today* **41**, 207–219 (1998).

[51] Ledoux M. J., Pham-Huu C., Guille J., Dunlop H., Hantzer S., Marin S., High specific surface area carbides of silicon and transition metals for catalysis, *Catalysis Today* **15**, 263–284 (1992).

[52] Pesant L., Elaboration d'un nouveau systeme catalytique a base de carbure de silicium (ß-SiC) pour la combustion des suies issues des automobiles moteur Diesel [Development of a new catalytic system based on silicon carbide (ß-SiC) for the combustion of soot from diesel automobiles], *Strasbourg 01*, Strasbourg, (2005).

[53] Bensaid S., Marchisio D. L., Fino D., Saracco G., Specchia V., Modelling of diesel particulate filtration in wall-flow traps, *Chem.Eng. J.* **154**, 211–218 (2009).

[54] Soldati A., Campolo M., Sbrizzai F., Modeling nano-particle deposition in diesel engine filters, *Chem.Eng. J.* **65**, 6443–6451 (2010).

[55] Glowinski R. and Wachs A., On the numerical simulation of viscoplastic fluid flow, *In Handbook of numerical analysis* **16**, 483–717 (2011).

[56] Balmforth N. J., Frigaard I.A., Ovarlez G., Yielding to stress: recent developments in viscoplastic fluid mechanics, *Annual Review Fluid Mech.* **46**, 121–146 (2014).

[57] Turan O., Poole R. J., Chakraborty N., Aspect ratio effects in laminar natural convection of bingham fluids in rectangular enclosures with differentially heated side walls, *J. Non-Newt. Fluid Mech.* **166**, 208–230 (2011).

[58] Turan O., Chakraborty N., Poole R. J., Laminar Rayleigh-Benard convection of yield stress fluids in a square enclosure, *J. Non-Newt. Fluid Mech.* **171**, 83–96 (2012).

[59] Hadidi N., Ould-Amer Y., Bennacer R., Bi-layered and inclined porous collector: Optimum heat and mass transfer, *Energy* **51**, 422–430 (2013).

[60] Amiri A. and Vafai K., Analysis of dispersion effects and non-thermal equilibrium, non-darcian, variable porosity incompressible flow through porous media, *Int. J. Heat Mass Transfer* **37**, 939–954 (1994).

[61] Peixinho J., Desaubry C., Lebouche M., Heat transfer of a non-newtonian fluid (carbopol aqueous solution) in transitional pipe flow, *Int. J. Heat Mass Transfer* **51**, 198–209 (2008).

[62] Mittal R. and Iaccarino G., Immersed boundary methods, *Annu. Rev. Fluid Mech* **37**, 239–261 (2005).

[63] Colella P., Graves D. T., Keen B. J., Modiano D., A cartesian grid embedded boundary method for hyperbolic conservation laws, *J. Computational Phys.* **211**, 347–366 (2006).

[64] Patankar S., *Numerical heat transfer and fluid flow*, J.CRC Press, 1980.

[65] Ragui K., Boutra A., Bennacer R., Benkahla Y. K., Progress on numerical simulation of yield stress fluid flows (part I): Correlating thermosolutal coefficients of bingham plastics within a porous annulus of a circular shape, *Int. J. Heat Mass Transfer* **126**, 72–94 (2018).

[66] Ragui K., Boutra A., Bennacer R., Benkahla Y. K., Critical dimension of a circular heat and solute source for an optimum transfer within square porous enclosures, *Energy Procedia* **139**, 817–823 (2017).

[67] Ragui K., Boutra A., Benkahla Y. K., Labsi N., Feddaoui M., TiO2-water nanouid within a tilted triangular enclosure including a square heater: Optimum heat transfer, *Mech. Industry* **17**, 612-622 (2016).

[68] Ragui K., Boutra A., Benkahla Y. K., Impact of square fuel assemblies arrangement on fluid flow and heat transfer enhancement inside water filled enclosures, *Int. Letters Chem. Phys. Astro.* **64**, 144 (2016).

[69] Ragui K., Boutra A., Benkahla Y. K., The impact of cylindrical fuel assemblies arrangement on the heat and mass transfers into an horizontal porous circular cylinder, *Int. Letters Chem. Phys. Astro.* **58**, 66 (2015).

[70] Ragui K., Benkahla Y. K., Labsi N., Boutra A., Natural convection heat transfer in a differentially heated enclosure with adiabatic partitions and filled with a bingham fluid, *Heat Transfer Research* **46** (2015).

[71] Boutra A., Ragui K., Labsi N., Benkahla Y. K., Free convection enhancement within a nanofluid filled enclosure with square heaters, *Int. J. Heat Tech.* **35**, 447–458 (2017).

[72] Boutra A., Ragui K., Labsi N., Bennacer R., Benkahla Y. K., Natural convection heat transfer of a nanofluid into a cubical enclosure: Lattice Boltzmann investigation, *Arab. J. Sci. Eng* **41**, 1969–1980 (2016).

[73] Ouyahia S. E., Benkahla Y. K., Labsi N., Numerical study of the hydrodynamic and thermal proprieties of titanium dioxide nanofluids trapped in a triangular geometry, *Arab. J. Sci. Eng* **41**, 1995–2009 (2016).

[74] Boutra A., Ragui K., Bennacer R., Benkahla Y. K., Three-dimensional fluid flow simulation into a rectangular channel with partitions using the lattice-boltzmann method, *Europ. Phys. J. Appl. Phys.* **74**, 24612 (2016).

[75] Boutra A., Ragui K., Benkahla Y. K., Numerical study of mixed convection heat transfer in a lid-driven cavity fillled with a nanofluid, *Mech.Industry* **16**, 505 (2015).

[76] Bishop E. H., Carley C. T., Powe R. E., Natural convective oscillatory flow in cylindrical annuli, *Int. J. Heat Mass Transfer* **11**, 1741–1752 (1968).

[77] Grigull U., Natural convection in horizontal cylindrical annuli, In: *Proc. 3rd Int. Heat Transfer Conf.*, 182–195 (1966).

[78] Trevisan O. V. and Bejan A., Natural convection with combined heat and mass transfer buoyancy effects in a porous medium, *Int. J. Heat Mass Transfer* **28**, 1597–1611 (1985).

[79] Beghein C., Haghighat F., Allard F., Numerical study of double-diffusive natural convection in a square cavity, *Int. J. Heat Mass Transfer* **5**, 833–846 (1992).

[80] Gobin D. and Bennacer R., Cooperating thermosolutal convection in enclosuresii. heat transfer and flow structure, *Int. J. Heat Mass Transfer* **39**, 2683–2697 (1996).

[81] Caltagirone J. P., Thermo-convective instabilities in a porous medium bounded by two concentric horizontal cylinders, *J. Fluid. Mech.* **76**, 337–362 (1976).

[82] Himasekhar K. and Bau H. H., Two-dimensional bifurcation phenomena in thermal convection in horizontal, concentric annuli containing saturated porous media, *J. Fluid. Mech.* **187**, 267–300 (1988).

[83] Benziane M. M., Abdul-Wahab S. A., Benaicha M., Belhadri M., Investigating the rheological properties of light crude oil and the characteristics of its emulsions in order to improve pipeline flow, *Fuel* **95**, 97–107 (2012).

[84] Benziane M. M., Abdelkrim Liazid, Bonneau O., Modelisation de lcoulement du ptrole brut dans les pipe-lines [Modeling of the flow of crude oil in pipelines], *Trans. Canadian Soc. Mech. Eng.* **32**, 09–22 (2002).

[85] Benziane M. M. and Zahloul H., Rheological behaviors of crude oil in the presence of water, *World Academy Sci. Eng. Tech. Int. J. Mech. Aeros. Industrial Mech. Manufac. Eng.* **07**, 521–525 (2013).

[86] Labonia G. and Guj G., Natural convection in a horizontal concentric cylindrical annulus: oscillatory flow and transition to chaos, *J. Fluid. Mech.* **375**, 179–202 (1998).

[87] Ohno Y., Enhanced heat transfer during oscillatory flow in annular channels, *Trans. JSME* **70**, 2612–2619 (2004).

[88] Dyakova V. and Polezhaev D., Oscillatory and steady flows in the annular fluid layer inside a rotating cylinder, *Shock Vib.* **2016**, (2016).

[89] Petrone G., Chnier E., Lauriat G., Stability analysis of natural convective flows in horizontal annuli: effects of the axial and radial aspect ratios, *Phys. Fluids* **18**, 104107 (2006).

In: Heat Conduction
Editors: J. Hristov and R. Bennacer

ISBN: 978-1-53614-673-8

Chapter 6

HYGROTHERMAL TRANSFERS IN POROUS MEDIA

Kamilia Abahri* and Xiaoyan Ma
LMT-ENS Cachan, CNRS, Paris-Saclay University,
Président Wilson, Cachan, France

Abstract

This first chapter is a synthesis of the current state of knowledge of the physical phenomena involved in the transfer of heat, air and moisture in porous media, mainly porous construction materials. The principal quantities characterizing the porous media are thus explained initially. Mechanisms involved in hygrothermal transfers are discussed. In a second step, a brief overview on transfer driving potential in porous media is mentioned.

Keywords: porous construction media, hygrothermal transfer

1. Preliminary Considerations on Porous Media-Projection on Materials

1.1. Porous Media, Status of Transfer

A porous media is a rigid solid of complex shape, inside which there are cavities or pores interconnected by channels. These pores may contain one or more fluid

*Corresponding Author Email: kamilia.abahri@lmt.ens-cachan.fr.

phases (liquid, vapor, gas) capable of flowing and exchanging with the solid matrix for matter and energy.

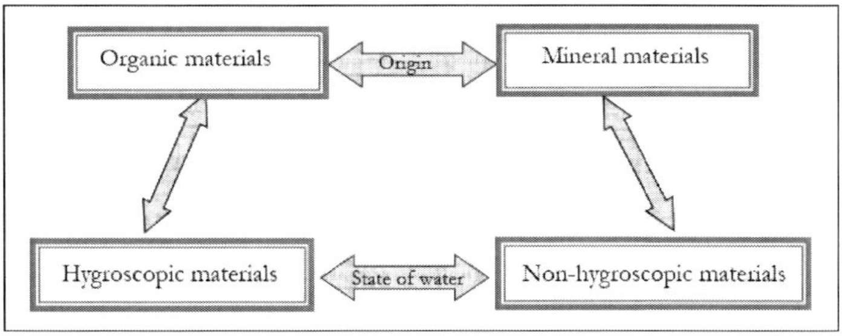

Figure 1. Classification of porous building materials.

There are several families of materials. Depending on the origin or the characteristic to water, four types can be cited in Fig. 1. Among them, we distinguish the family of *hygroscopic* materials that is attributed to materials with the ability to fix a measurable amount of moisture from the surrounding air. Experimentally, the hygroscopicity of the materials is revealed by the variation of their mass when they are subjected to humid ambient. Wood-based materials like wood fiber, cellulose wadding are examples of this type of material. They are reactive to water, but degrade for high water content. Other materials such as cellular concrete and calcium silicate can withstand high levels of water but are less reactive to water.

In addition, there are other types of classification of materials based on their mode of application such as thermal insulation, which are one of the most cost-effective ways of saving energy in construction. Indeed, a powerful thermal insulation must meet withe the four main characteristics:

- the lightness and the non-hygroscopicity in order to preserve in time its insulating qualities,
- to be an obstacle to heat transfer, having a low thermal conductivity,
- with a good mechanical resistance,
- the possibility of being laid with a sufficient thickness, so as to obtain a significant thermal resistance.

1.2. Parameters Characterizing the Porous Media

1.2.1. Representative Elementary Volume (R.E.V.)

In the study of transfer phenomena in porous media, two scales are necessary for the description of phenomena [1]: One is the pore scale, or microscopic, within which the local quantities can vary widely. In general, this scale is associated with the average pore diameter, i.e., d in $[m]$. The other is the scale of the porous media, or macroscopic, characteristic of significant variations of these same quantities, defined by an average value on a certain volume of porous media, commonly called Elementary Volume Representative and noted *R.E.V.* This macroscopic scale is associated with the geometric dimension of the medium, i.e., L in $[m]$.

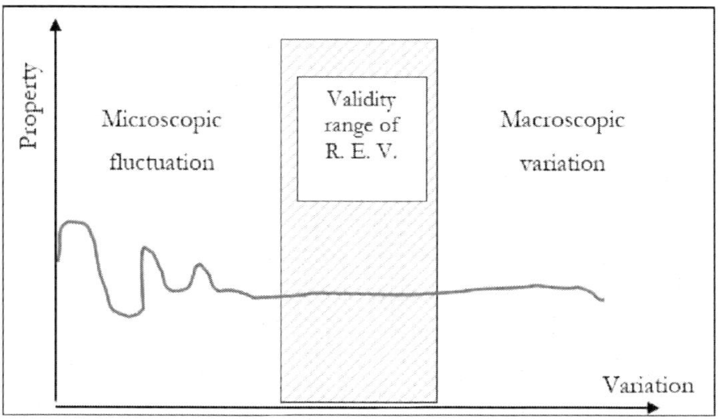

Figure 2. Definition of the representative elementary volume, on which the property P is averaged.

The macroscopic quantities characterizing the porous media are thus averages of the corresponding microscopic magnitudes, taken on a domain whose size is large in comparison to the scale of their microscopic fluctuations, but small compared to that of the variations at the macroscopic scale (Fig. 2). Thus, l designating the characteristic dimension of R.E.V., must satisfy the following inequality: $d < l < L$. To carry out the change of scale (i.e., passage from the pore scale to the macroscopic scale), several methods have been proposed, among which are the averaging method [2], and the method of periodic homogenization [3], [4], [5], [6], [7].

1.2.2. Porosity

The porosity of a porous media, denoted $\varepsilon[-]$ is defined as the ratio of the volume of the cavities to the volume occupied by the porous media:

$$\varepsilon = \frac{V_l + V_g}{V_t} \tag{1}$$

In which V_l $[m^3]$, V_g $[m^3]$, V_t $[m^3]$ are the volume of the liquid phase, the volume of the gas phase, and total volume, respectively.

1.2.3. Intrinsic and Apparent Permeability

The permeability of a material characterizes its ability to be passed through by a fluid (vapor, liquid, gas) under a total pressure gradient (according to Darcy's law). The flows are at a fairly slow rate; the flow regime is laminar; the viscous forces are predominant and the Darcy's law is applicable: the pressure gradients are proportional to the flow velocity in the pores. This relationship of proportionality is maintained by averaging the flow rate and total pressure gradients on a large volume in front of the pore size. For a specimen with a section of $S[m^2]$, length of $L[m]$, the volume flow of $Q_v[m^3/s]$, the following relation is satisfied:

$$\frac{Q_v}{S} = \frac{k}{\mu}\frac{\Delta p}{L} \tag{2}$$

The ratio of the total volume flow per unit area of the materia, Q_v/S, represents the filtration rate of the fluid $V[m/s]$, which corresponds to the average speed of the fluid taken by a R.E.V. $\mu[kg/m \cdot s]$ is the dynamic viscosity of the fluid (water in this case) and the proportionality constant $k[m^2]$ is the permeability coefficient of the porous media.

The apparent permeability (k/μ) is homogeneous to a surface and it depends on the porosity and geometry of the solid matrix. It is obtained for a given pressure gradient in the case of a compressible or incompressible fluid and depends on this gradient. The intrinsic permeability $k[m^2]$ is the property of a material to transmit a fluid under pressure. It is independent of the characteristics of the fluid. Several evolutions of the permeability as a function of the porosity and a characteristic dimension of the porous matrix at the pore scale have been proposed. The different expressions giving the permeability depend on the geometry of the porous media.

Kozeny Carman's law (1937) provides an estimate of the permeability $k[m^2]$ for an unconsolidated porous media consisting of identical elements of simple geometry:

$$k = \frac{d^2\varepsilon^2}{36 \cdot C_0(1-\varepsilon)^2} \tag{3}$$

In which, $d[m]$ denotes a characteristic dimension of the elements constituting the porous matrix. $C_0[-]$ is a constant depending on the grain shape ($3.6 < C_0 < 5$) and in particular $C_0 = 4.84$ for spherical grains. $\varepsilon[-]$ is the porosity of the material. The modeling of pores by parallel capillary veins of width G and distant from T, leads to a permeability:

$$k = \frac{G^3}{12(G+T)} = \frac{G^2}{12}\varepsilon \qquad \varepsilon = \frac{G}{G+T} \tag{4}$$

The modeling of the porous matrix as a bundle of parallel capillary tubes leads to the following expression of the permeability:

$$k = \varepsilon\frac{d^2}{32} \qquad (\varepsilon = n(\pi d^2/4)) \tag{5}$$

Where n is the number of capillaries per unit area perpendicular to the direction of flow and d the diameter of the capillaries.

1.3. Interaction of Moisture between the Material and Air

1.3.1. Humid Air

A porous material is considered dry if it contains no water or chemically bound water. When in contact with moist air, mixed in varying proportions of dry air and water vapor, hygroscopic materials tend to change their initial water state. Humid air is perfectly defined by its temperature and the body of water it contains. Its moisture content, called absolute humidity, remains constant even if the air temperature varies with the condition that it remains higher than the dew point temperature. Moisture in the air is mainly present in vapor form and the maximum amount of vapor in the air depends strongly on the temperature.

1.3.2. Dew Temperature

The dew temperature or dew point temperature is the temperature at which the partial pressure of the vapor in the humid air is equal to the saturation vapor pressure, causing the condensation of the water vapor.

1.3.3. Relative Humidity

If the air is not saturated with moisture, the absolute humidity of the air is lower than that obtained at saturation. As a result, the relative humidity quantity $\varphi[\%]$ can be defined. It is the ratio between the partial pressure of water vapor $p_v[Pa]$ and the saturation vapor pressure $p_{sat}[Pa]$ expressed at the same temperature:

$$\varphi = 100\frac{p_v}{p_{sat}} \tag{6}$$

This parameter is also frequently named as RH (Relative Humidity), in which $p_{sat}[Pa]$ is the saturation vapor pressure. It is defined as the equilibrium pressure between liquid water and water vapor. It is an increasing function of the thermodynamic variable - the temperature $T[K]$. Numerous parameterizations exist, a valid empirical expression for $0 < T < 80^0C$ with a precision of $\pm 0.15\%$.

$$p_{sat}(T) = exp(23.5771 - \frac{4042.9}{T - 37.58}) \tag{7}$$

Where $T[K]$ is the absolute temperature, and $p_{sat}[Pa]$ is the saturated vapor pressure.

1.4. Moisture Content

Depending on environmental conditions, moisture in building materials can occur in micropores in the form of solid, liquid or vapor. The water content (or moisture content) of a material is the ratio of the weight of water contained in this material to the weight of the same dry material. It can also be defined as the weight of water contained per unit weight of dry material. This property can be evaluated by the weight or volume ratio.

To characterize a porous material in terms of transfers, it is important to define a number of physical parameters. At the macroscopic level, the complete description of such a media therefore presupposes the description of each of its phases and their relative proportions:

- A solid phase which constitutes the unconsolidated granular skeleton, characterized by its mass $m_0[kg]$ and its volume $V_0[m^3]$.
- A liquid phase which is in this case of water, mass $m_l[kg]$ and volume$V_l[m^3]$.
- A gaseous phase, mixing air and water vapor. This last mass $m_g[kg]$ occupies a volume $V_g[m^3]$.

For a medium of total mass $m = \sum_i m_i(i = 0$ for the solid phase,$i = v$ for the vapor phase, $i = l$ for the liquid phase, $i = g$ for the gaseous phase and $i = a$ for dry air) occupying a total volume $V_t = \sum_i m_i$, a certain number of quantities can be defined: The content of each substance *i* can be:

- Mass ω_i in $[wt\%]$:

$$\omega_i = \frac{m_i}{m_0} \tag{8}$$

- Mass per unit volume u in $[kg/m^3]$:

$$u_i = \frac{m_i}{V_t} = \rho_0 \omega_i \tag{9}$$

- Volumic Θ in $[vol\%]$:

$$\Theta_i = \frac{V_i}{V_t} \tag{10}$$

- The degree of water saturation:

$$S_l = \frac{V_e}{(V_t - V_0)} \tag{11}$$

Where $m_i[kg]$ is the mass of water in the sample, $m_0[kg]$ is the dry mass of the sample, $V_t[m^3]$ is the total volume of the sample, $\rho_0[kg/m^3]$ is the dry density of the sample, $V_e[m^3]$ is the volume of water present in the porous sample backbone. These different quantities are connected to each other by the following relations:

$$\varepsilon S_l = \Theta_l = \frac{\rho_s}{\rho_l} u_l \tag{12}$$

Where $\rho_l[kg/m^3]$ is the density of liquid water, and $\varepsilon[-]$ is the porosity of the porous media.

1.4.1. Total Pressure

From the thermodynamic point of view, in a perfect gas mixture, the total pressure $p[Pa]$ is the sum of the partial pressure of eeach constituent of the mixture (Dalton relation). Considering moist air as the mixture of dry air and water vapor and assuming that these two gases follow the behavior of a perfect gas, the total pressure of moist air is the sum of the partial pressure water vapor $p_v[Pa]$ and the partial pressure of dry air $p_a[Pa]$:

$$p = p_v + p_a \tag{13}$$

On the other hand, certain standards related the hygrothermal performance to components and building walls [8], affecting the humidities and the pressures of the indoor air and the outside air of the walls of the building, which define indoor hygrothermal load classes. This classification is related to the type of a local activity (residential building, tertiary or secondary activity, swimming pool, etc.). In this same standard, it is proposed to obtain the humidity of the indoor air by one of the two following formulas:

$$p_{int} = p_{ext} + \Delta p \tag{14}$$

$$\varphi_{int} = \varphi_{ext} + \Delta\varphi \tag{15}$$

where $\Delta p[Pa]$ is the excess of internal vapor pressure, $\Delta\varphi[-]$ is the excess of indoor humidity, $p_{int}[Pa]$ is the pressure in the indoor environment, $\varphi_{int}[\%]$ is the relative humidity in the indoor environment. The values of Δp and $\Delta\varphi$ are fixed in relation to the intended use of the building which are classified hygro-thermally; so they can be either storage areas, offices, businesses, low-occupancy housing, high-occupancy housing, kitchens, special buildings (laundry, pool, brewery). Indeed, the walls of buildings are generally exposed to low gradients of total pressure. These pressure differences come from ventilation, wind, draft or air density variation, etc. These mechanisms can be combined to determine the direction and amplitude of air pressure. The forces produced by each of them are generally known. In addition, air infiltration is responsible for many problems such as condensation, ice formation on exterior surfaces and lack of control of temperature and humidity.

2. Method of Confining Moisture in Porous Media

Two fundamental mechanisms help to confine a fluid in the porous media. These are respectively adsorption and capillarity. However, it is necessary to add, when the medium is likely to react chemically with moisture, other chemical or osmotic binding mechanisms. These last modes of connection are not involved in this study. They will not be mentioned later in the manuscript.

2.1. Adsorbed Water

According to the BET theory [9], which is a generalization of an older mono-molecular theory [10], the water molecules are deposited on the solid in several superimposed layers, and each time covering a surface smaller than that of the previous layer. The binding energy from the second layer is equal to the latent heat of vaporization. The volumetric water content $\theta[vol\%]$ is determined by the available solid surface area per unit volume, the number of deposited layers and the thickness of a single-molecule layer. The equilibrium of the exchanges of water molecules between the vapor above the surface and the condensed phase gives the following relation:

$$\theta = \frac{C \cdot e_m \cdot S_0 \cdot \varphi}{(1-\varphi)(1+(C-1)\varphi)} \tag{16}$$

in which, $e_m[m]$ is the thickness of a molecule layer, $S_0[m2]$ is the specific surface area of the solid (area per unit volume), $\varphi[\%]$ is the relative humidity defined previously, and $C[-]$ a coefficient dependent on the temperature and binding energy of the molecules to the solid in the form of:

$$C = exp(\frac{E_l - L_m}{RT}) \tag{17}$$

where, $E_l[J/mol]$ is the binding energy of the first layer of molecules, $L_m[J/mol]$ is the latent heat of vaporization, $R[J/mol]$ is perfect gas constant, and $T[K]$ the absolute temperature. A building material can theoretically absorb moisture until all these pores are filled with water. On the other hand, this fact does not provide any information as to its actual moisture storage capacity. It is therefore important to establish a connection between the water content of the building materials and the ambient conditions.

The graphical representation of Figure 3 shows the measurements of the variation of the water content of the sample as a function of the relative humidity. These curves are characteristic of a given material and a given temperature. They are called isothermal sorption. Experimentally, they are obtained starting from a dry sample while following the amount of water vapor stored by the porous material for the set of state of equilibrium corresponding to relative humidity between 0 and 100 $\%$. The general shape of the adsorption isotherm curves for different hygroscopic materials is not the same (Figure 3). According to this figure, wood is the most hygroscopic material compared to brick and other types of concrete because it has much higher moisture content in the hygroscopic area.

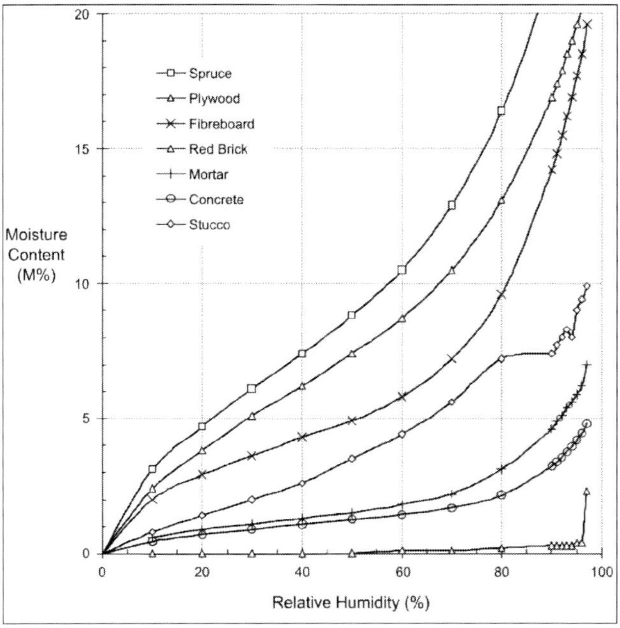

Figure 3. Sorption of moisture for different materials[11].

2.2. Capillary Water

When wetting a porous material with a water content $u[kg/m^3]$ higher than the maximum water content $u_c[kg/m^3]$, liquid menisci will be formed between the pores. This phenomenon results from the particular properties of wettability of water relative to the solids constituting the porous matrix. This leads to the the formation of curved interface between the wetting fluid (water) and the gaseous non-wetting fluid (moist air). The analysis of the equilibrium conditions of such an interface shows the existence on both sides of a positive pressure difference between moist air and water, called capillary pressure p_c presented as:

$$p_c = p_a - p_l \tag{18}$$

where $p_a[Pa]$ and $p_l[Pa]$ are the air pressure and the liquid water, respectively.

This measurable pressure difference $p_c[Pa]$ is theoretically expressed according to the Laplace law as a function of the surface tension $\sigma(T)$ of the air-water pair, the main radius of curvature $R_1[m]$ and $R_2[m]$ of the interface and the wetting angle defined by the tangents between the solid phase and the interface from the fluid-solid contact point:

$$p_c = \sigma \cdot cos\theta(\frac{1}{R_1} + \frac{1}{R_2}) \tag{19}$$

For a capillary of circular cross-section of radius r, the preceding expression becomes:

$$p_c = \sigma \cdot cos\theta / r \tag{20}$$

Where $\sigma[kg/m]$ is the water-vapor surface tension.

The capillary pressure can be expressed in the height of water which is called suction $\psi[m]$:

$$\psi = -\frac{p_c}{\rho_l g} = \psi(T, u) \tag{21}$$

Where $\rho_l[kg/m^3]$ is the density of water, and $g[m/s^2]$ acceleration of gravity.

The suction varies according to the water content $u[kg/m^3]$. It vanishes for a water content equal to the saturation water content u_{max}, for which all the pores are filled, and increases significantly in the vicinity of the hygroscopic domain. It should however be noted that there are hysteresis phenomena, the equilibrium profile in drainage being different from the equilibrium profile in humidification. The total hygroscopic suction is often expressed by the suction pressure obtained by Kelvin's law, the expression of which is as follows:

$$p_c = \frac{RT\rho_l}{M} ln(\varphi) \tag{22}$$

where, $R[J/mol \cdot K]$ is the molar constant of perfect gases ($R = 8.31443 J/mol \cdot K$), $\rho_l[kg/m^3]$ is the density of the water, $M[kg/mol]$ is the molar mass of water, $\varphi[\%]$ is the relative humidity, and $T[K]$ is the thermodynamic temperature. This expression shows a match between ψ and φ. This will determine the hygroscopic suction (for humidities below 95 %).

However, there is a part of the interstitial water called free water because it can flow between the solid particles due to pressure differences generated by various actions such as gravity, the application of a charge to the surface or in soil, desiccation or transformation of water into ice.

2.3. The Hysteresis

The relationship between water content and relative humidity, in experimental practice, is not unique but depends on the direction of the evolution of relative humidity RH, presenting a *hysteresis*. Different forms of loops can be schematized. They are grouped in four classifications (Figure 4) [12]: the two loops of hysteresis noted H_1 and H_2 appear respectively when the material has a very narrow distribution of mesopores and in the case of materials having mesopores in intercommunication. The hysteresis branch H_3 is described in the case where the material forms aggregates. It can be attributed to capillary condensation taking place in a non-rigid texture and is not characteristic of a defined mesoporosity. The last branch of hysteresis is often observed with microporous adsorbents having sheets bound together in a more or less rigid manner and between which capillary condensation may occur.

Two main causes of hysteresis have been expressed as: the first interpretation refers to the difference between the radius of curvature of a cylindrical surface and a spherical surface [13]. During the sorption, the evolution of the fluid-vapor interface is done first from a cylindrical surface (the walls of the pore). In desorption, the evolution is made from a spherical meniscus. The radius of curvature of a spherical surface being half that of a cylindrical surface of the same radius. This is sometimes called the *boot problem*.

The second cause of hysteresis is the accessibility of certain pores by pores of smaller diameter, in configuration *ink bottle*. In sorption the filling of the access pores prevents the filling of the pores of greater diameter. The air remains trapped as soon as the access grooves are filled. In desorption, emptying the larger diameter pore starts only when the relative humidity drops low enough to empty the pore of smaller diameter. In the intermediate states, equilibrium

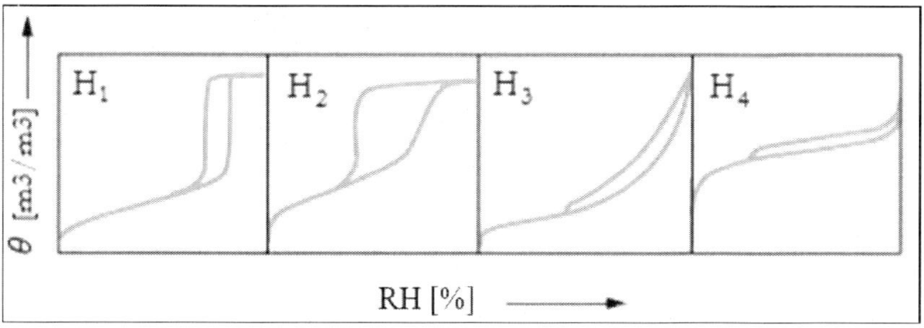

Figure 4. Type of hysteresis (classification given by IUPAC).

situations can be established at the same relative humidity for different volumes of liquid and therefore different water contents. One study explains the hysteresis differently: in the case of adsorption, all the pores are initially opened and the water condenses firstly in the small pores by the effect of the capillarity forces [14]. On the other hand, the condensation in the macro-pores sites will take place only at high relative humidities, which depend on the pore diameter too. These different behaviors during adsorption and desorption cause the effect of hysteresis. There are several models in the literature that explain this phenomenon, based on the nature of the adsorption and desorption curves that are usually off balance. In addition, the hysteresis can be directly related to the shape of the pores or their structure.

3. Quantitative Analysis of Moisture Transfer Phenomena

In the field of civil engineering, moisture in the material is the most often considered as water vapor. The vapor transfer in porous building materials is broadly classified into three modes, which is presented representively in the following.

3.1. Molecular Diffusion

Molecular diffusion or Fick diffusion is dominated in pores at a larger radius $10^{-6}m$ [15], where the transport of water vapor is governed by the collision between particles or molecules (Fig.4a). In this type of diffusion, there is no interaction with the matrix of the material. In the open air, the diffusion coefficient proposed the following equation [16] :

$$D_a = c(\frac{p_0}{p})(\frac{T}{T_0})^n \quad (23)$$

Where the reference pressure is $p_0 = 1.101325 \times 10^5 Pa$, the reference temperature $T_0 = 273.15K$, the factor $n = 1.88$ and $c = 2.17 \times 10^5 m^2/s$.

3.2. KNUDSEN Transport and Surface Scattering

In this type of diffusion (also called effusion), the gap between the pores is considered as a rough surface that imposes different tortuosity of the adsorbed diffusion pathway. The movement of molecules is governed by molecular velocity as well as collisions with other molecules (Figure 4b). The collision between the diffusion molecules creates resistance for diffusion. If the diffusion of gas takes place in a very fine pore, at a low pressure, the average free path of the molecules must be greater than the diameter of passage. The rate of matter transfer by diffusion of a species is governed by its molecular velocity, the diameter of passage and the pressure gradient. This is what is called *Knudsen's broadcast*. This phenomenon becomes important if the pore size is less than $10^{-8}m$. Some models have been proposed to describe and quantify the diffusion of Knudsen: A simple approach based on the kinetic theory of gases gives the following expression for the diffusivity of Knudsen given by [18]:

$$D_k = \frac{2}{3}\sqrt{\frac{8RT}{\pi M}} \cdot r_p \quad (24)$$

In which, $r_p[m]$ is the radius of passage of capillaries, $T[K]$ is the temperature, $M[kg/mol]$ molecular weight, and $R[J/mol{\cdot}K]$ is the molar constant of perfect gases ($R = 8.31443 J/mol \cdot k$).

Other important diffusion phenomena called *surface diffusion* are worth mentioned. It involves the transport of molecules absorbed on the surface under the presence of the water vapor gradient (Figure 4c). If the ratio of the amount adsorbed to that required to form a monolayer is less than unity, some active sites remain empty. An adsorbed molecule tends to migrate to an empty adjacent site if it has enough energy to cross the energy barrier. Since the active sites are discrete, the migration of the surface diffusion is visualized to be reproduced by a *jump*. The flux due to surface scattering is given by Fick's law [17], [18].

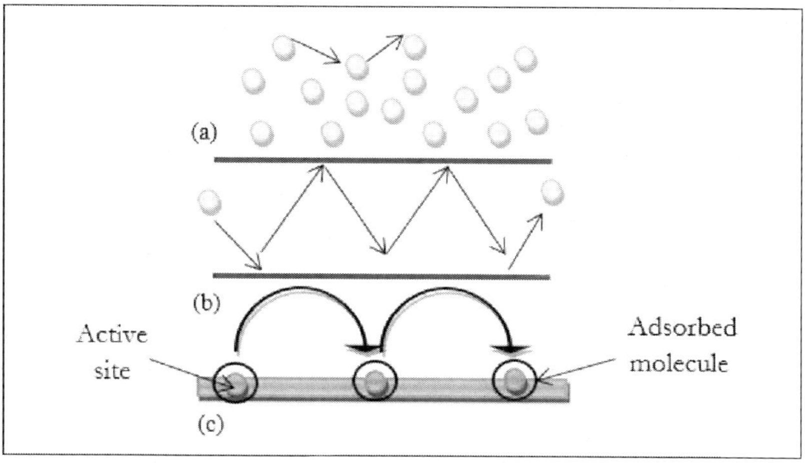

Figure 5. Illustrative of (a) diffusion in a continuum, (b) diffusion of Knudsen and (c) surface diffusion.

$$j_{vs} = D_{vs} \bigtriangledown u_v \tag{25}$$

In which, $D_{vs}[m^2/s]$ area diffusion coefficient, $j_{vs}[mol/m \cdot s]$ is the diffusion flux at the surface, and $u_v[kg/m^3]$ the vapor content.

However, surface diffusion as well as capillary conduction can occur only in a liquid phase. It is indicated that this kind of diffusion is usually a part of liquid transport and not a part of vapor diffusion [17].

3.3. Thermal Diffusion

Due to the difference in density between dry air and vapor molecules, it makes the passage of water vapor from cold to hot. Therefore, it results in a transport as a function of the temperature gradient. Under isothermal conditions, the theory of vapor transport in porous materials is initially based on Fick's law (ion diffusion in water). Fick's law is generally given for the vapor content as follows:

$$j_v = D_v \bigtriangledown C_s \tag{26}$$

In which, $C_s[kmol/m^2]$ is the concentration at the surface of the adsorbed molecules. This law is adopted in the physical construction to describe the diffusion of water vapor in porous materials. Otherwise, the diffusion coefficient of water vapor in the air can be defined by the following relationship [17].

$$\delta_a = 2 \times 10^{-7} T^{0.18} / p \tag{27}$$

In which, $p[Pa]$ is the ambient air pressure, $T[K]$ is the temperature, and $\delta_a[m^2/s]$ is the diffusion coefficient of water vapor in the air.

4. Main Models of Water Transfer

4.1. Darcy's Law (1856)

This law is widely used when it comes to transfers in porous media. The filtration rate $V[m^2/s]$, macroscopic velocity, is distinguished from the interstitial velocity, denoted $V_p[m^2/s]$, which corresponds to an average velocity of the fluid in the pores. This actual velocity $V_p[m^2/s]$ inside the pores can be much greater than V, since only a portion of the total volume of the porous material is available for the transport of the fluid. The validity limit of Darcy's law is defined by the Reynolds number *Re* which depends on the interstitial velocity V_p, a characteristic length of the pore size $d[m]$:

$$Re = \rho V_p \frac{d}{\mu} \tag{28}$$

One study distinguishes three flow regimes according to the value of the pore Reynolds number [19]. Another using velocity measurements, distinguish several flow regimes that can be classified as follows [20]:

- For pore Reynolds numbers less than 1, the flow regime is laminar, viscous forces are predominant and Darcy's law is applicable.
- When *Re* is between 1 and 10, the boundary layers at the solid walls of the pores develop. Outside these boundary layers appears an inertial flow regime. The forces of inertia then become important and there is no more proportionality between the pressure gradient and the filtration speed: Darcy's law is no longer applicable. This stationary laminar flow regime persists to a Reynolds number of about 150.

- For pore Reynolds numbers between 150 and 300, an unsteady laminar flow regime develops. This regime is initially oscillatory resulting from the superposition of several distinct modes of period and amplitude.
- Beyond *Re* value of 300, the flow regime becomes chaotic, and has the characteristics of a turbulent flow.

4.2. The Empirical Model of Brinkman

When the porosity of the medium becomes important, Darcy's law becomes insufficient to describe the fluid flow within the porous media. Thus, it is important to take into account the viscous diffusion effects created by the friction at the walls, the induced viscous forces then becoming the same order of magnitude as the resistance forces created by the porous matrix. Brinkman was the first author to add a term representing viscous constraints to Darcy's law [21]. This model was taken over and improved later by other authors. Taking inspiration from the Stokes formula, Brinkman proposed a modification of Darcy's law by adding the term $\mu^{'} \bigtriangledown^{2} V$ or $\mu^{'}$ is the effective viscosity of the porous media, which leads to the equation:

$$\bigtriangledown p = \rho \frac{\mu^{'}}{K} V + \mu^{'} \bigtriangledown^{2} V \tag{29}$$

where $K[m^2]$ is the permeability of the porous media, $V[m^2/s]$ is corresponding to the average fluid velocity taken by a representative elementary volume, $\mu[kg/m \cdot s]$ is the dynamic viscosity of the fluid, $\mu^{'}[kg/m \cdot s]$ is the effective viscosity given by the Brinkman model [22].

4.3. Model with Driving Potential of Steam Pressure/Capillary Pressure

4.3.1. Pedersen Model

In order to evoke the discontinuity of water content as a driving potential in moisture transfer in multilayer materials, Pedersen chose the suction pressure as a capillary transfer potential[23]. The direct relationship between water content and equilibrium capillary suction pressure can be determined in the upper relative humidity range by pressure bearing measurements [17]. In the moisture adsorption region ($< 98\%$ relative humidity), the capillary suction pressure can be calculated as a function of the relative humidity according to the Kelvin law (Eq. 20). On the other hand, for non-capillary active materials, such as most thermal insulators, the capillary suction pressure can not be defined and no liquid transfer takes place. The mass balance equation thus obtained is as following:

$$\frac{\partial u}{\partial t} = \nabla(\delta_p \nabla p_v - k \cdot p_c \nabla (ln(p_c)) \tag{30}$$

where $u[kg/m^3]$ is the volume water content, $p_c[Pa]$ is the capillary pressure, $p_v[Pa]$ is the vapor pressure, $k[kg/m \cdot s \cdot Pa]$ is the hydraulic conductivity, $\delta_p[kg/m \cdot s \cdot Pa]$ is the vapor permeability.

This equation decomposes the model into two separate transport equations for the liquid phase and the vapor phase [23]. In his model, δ_p is a function of relative humidity. The capillary pressure gradient represents the liquid transfer and the use of $ln(p_c)$, which is only introduced for numerical reasons here.

4.4. Liquid Diffusion Model

4.4.1. Richards Model

Some research focuses on the mechanisms of water transport in partially saturated porous media [24]. Although Darcy's equation was originally designed for saturated media flows, it was extended to unsaturated flow, stating that the proportionality constant k, called hydraulic conductivity, is a function of soil moisture content [24].

$$\frac{\partial \theta}{\partial t} = \frac{\partial}{\partial z}[k(h)(\frac{\partial h}{\partial z} - 1)] \tag{31}$$

Where $\theta[vol\%]$ is the water content, $k[m/s]$ is the hydraulic conductivity, and $h[m]$ is a relative pressure with respect to the atmospheric pressure of water expressed as a function of height. The pressure of the gas phase in the porous media is considered constant and equal to the atmospheric pressure. This is a limitation of this model especially in case of infiltration since the pressure of the air trapped in the ground can increase. This equation is fully defined if we know the hydraulic conductivity k and the capillary capacity C as a function of the water content θ and the pressure h. Richards also introduced a relative permeability dependent on θ and h. While the intrinsic permeability only depends on the geometric properties of the dry porous media. The relative permeability depends on the fluids present in the system on the one hand and the pressure of the latter on the other hand.

4.5. Water/Temperature Transfer Models

4.5.1. Model of Philip and De Vries (P.D.V)

Philip *et.al* have considered the transport of water vapor as a diffusive process governed by Fick's law, such as [16] :

$$j_v = -D_a \cdot \nu \cdot \tau \cdot a_v \bigtriangledown \rho_v \tag{32}$$

In which, j_v is the vapor density vector, D_a is the binary diffusion coefficient of the water vapor in the air, τ is the tortuosity related to the connectivity of the continuous paths through the pore space, a_v is the volumetric content in the gaseous phase, p_v is the density, and ν is a coefficient related to the diffusion of steam. During diffusion by condensation evaporation this coefficient is given by: $\nu = p^{'}/(p^{'} - p_\nu), p_v$ is the total pressure and the partial pressure of the water vapor. It should be noted that $D_a = 4.42 \times 10^4 \times T^{2.3}/p^{'}$ used for temperatures that do not exceed 70 °C. The gradient of the vapor density is a function of the temperature gradient and the water content:

$$\bigtriangledown \rho_v = \varphi \frac{d\rho_s}{dt} \bigtriangledown T + \frac{g\rho_v}{RT}\frac{dp_l}{d\Theta} \bigtriangledown \Theta \tag{33}$$

where $\rho_s[kg/m^3]$ is the dry density of the material, and $p_l[Pa]$ is the liquid pressure.

The injection of Eq.33 into Eq.32 leads to the following vapor flow expression:

$$j_v/\rho_l = -D_{Tv} \bigtriangledown T - D_{\Theta v} \bigtriangledown \Theta \tag{34}$$

In which, $D_{Tv}[m^2/s]$ is the non-isothermal diffusion coefficient of water vapor, $D_{\Theta v}[m^2/s]$ is the isothermal diffusion coefficient of water vapor.

These diffusion coefficients are deduced from the sorption isotherms or the capillary characteristic of the material [25]. The liquid transfer derives from Darcy's law. It appears as the macroscopic result of the laminar viscous flow of the fluid in the complex network of microscopic conduct, this allows to write:

$$j_l/\rho_l = -D_{Tl} \bigtriangledown T - D_{\Theta l} \bigtriangledown \Theta - k_i \tag{35}$$

in which, $D_{Tl}[m^2/s]$ is the non-isothermal diffusion coefficient of liquid water, $D_{\Theta l}[m^2/s]$ is the isothermal diffusion coefficient of water vapor, $k_i[m/s]$ is the hydraulic conductivity. The general differential equations describing the transport of moisture in porous materials under thermohydric gradients are:

$$\frac{\partial \Theta}{\partial T} = \triangledown(D_T \triangledown T) + \triangledown(D_\Theta \triangledown \Theta) + \triangledown k_i \tag{36}$$

$$\rho c \frac{\partial T}{\partial t} = \triangledown(\lambda \triangledown T) + L_{vl} \triangledown (D_\Theta v \triangledown \Theta) \tag{37}$$

where, $D_T[m^2/s]$ is the non-isothermal diffusion coefficient of liquid water, $D_\Theta[m^2/s]$ is the isothermal diffusion coefficient of water vapor, $L_{vl}[J/kg]$ is the latent heat of vaporization, $c[J/kg \cdot K]$ is the heat capacity of the material, $\lambda[w/m \cdot K]$ and thermal conductivity.

The energy equation is only described by conduction heat transfer and phase change. The *P.D.V* model does not take into account the evolutions of the pressure in the material, it is generally used in the field of drying.

4.6. Total Pressure Driving Potential Models/Steam Content

Most transfer models available in the literature, including those cited above, have often been developed without integrating the total pressure gradient on the thermohydric behavior of materials. From the theoretical and experimental point of view, numerous indexes justify the need for such integration of this new transfer driving potential.

Among the few exsiting works have confirmed the possibility of incorporating the total pressure introduced by [26], [27]. They assumed that during an excessive period of drying, a total pressure gradient appears in the material. This total pressure gradient generates additional heat and moisture transport resulting from the movement of vapor and liquid infiltration into the material [28]. They showed that the total pressure gradient in the material appears as a result of evaporation and resistance of the vapor movement in the porous skeleton. In addition, the effect of atmospheric pressure on simultaneous heat and moisture transfer in hygroscopic textiles has been studied [29]. It has been confirmed that the total pressure gradient has a very significant effect on transfers by comparing two analytic cases of an analytical model developed.

A mathematical model was developed to predict transfers in a block of hollow brick material [30]. The phenomena taken into account in this model are:

water vapor transfer, liquid water transfer, evaporation (respectively condensation) and heat transfer. The vapor transfer is by diffusion/convection and the liquid flow is governed by Darcy's law. As for heat transfer, it is done by convection and conduction. Dry air is transferred by diffusion / convection. The main flows describing these transfers are given below [30]:

$$j_l = K(\nabla p_{suc} - \rho_l g) \tag{38}$$

$$j_v = -\delta_v \nabla p_v - \rho_v \frac{k_g k_{rg}}{\mu_g} \nabla p_g \tag{39}$$

$$j_a = -\delta_v \nabla p_v - \rho_a \frac{k_g k_{rg}}{\mu_g} \nabla p_g \tag{40}$$

$$j_q = \lambda \nabla T + j_l C_{pa} T + j_v L + j_v C_{pv} T \tag{41}$$

where, $j_l[kg/m^2 s]$ is the density of the liquid flow, $j_v[kg/m^2 s]$ is the density of the vapor flow, $j_a[kg/m^2 s]$ is the density of the dry air flow, $j_q[kg/m^2 s]$ is the density of flow of heat, $\delta_v[s]$ is the vapor permeability, $p_v[Pa]$ is the vapor partial pressure, $p_{suc}[Pa]$ is the suction pressure, $\rho_v[kg/m^3]$ is the density of the vapor, $\rho_v[kg/m^3]$ is the density of the liquid water, $K[s]$ is the liquid permeability, $k_g[m^2]$ is the absolute permeability, $k_{rg}[m^2]$ is the relative permeability, $\mu_g[Pa \cdot s]$ is the dynamic viscosity, $p_g[Pa]$ is the gas pressure, $\lambda[w/m \cdot K]$ is thermal conductivity, $C_{pa}[J/kg \cdot K]$ is the constant pressure heat capacity of dry air, $C_{pl}[J/kg \cdot K]$ is the heat capacity of liquid water, $C_{pv}[J/kg \cdot K]$ is the heat at constant vapor pressure, and $L[J/kg]$ is the latent heat of vaporization. Finally, the establishment of the equations of mass balances and heat makes it possible to obtain a system of differential equations:

$$\frac{\partial u}{\partial t} = -\nabla (j_v + j_l) \tag{42}$$

$$\frac{\partial \rho_a}{\partial t} = -\nabla (j_a) \tag{43}$$

$$C_p \rho_s \frac{\partial T}{\partial t} = -\nabla q \tag{44}$$

where $\rho_s[kg/m3]$ is the dry density, $C_p[J/kg \cdot K]$ is the heat capacity of the material, $u[kg/m^3]$ is the moisture content by volume.

A mathematical model was also developed to predict coupled heat, moisture and air transfers [31]. The phenomena taken into account in this model are: water vapor transfer, liquid water transfer, evaporation (respectively condensation) and heat transfer. The vapor transfer is by diffusion/convection and the liquid flow is governed by Darcy's law. As for heat transfer, it is done by convection and conduction. Dry air is transferred by diffusion/convection. In addition, they took into consideration the phenomenon of thermodiffusion through the incorporation of a temperature gradient thus constituting an additional transfer engine. The phenomina can be discribed by the following equations.

$$\frac{\partial \omega}{\partial t} = div[D_m(\nabla \omega + \delta \nabla T + \alpha \nabla p)] \tag{45}$$

$$c\rho_s \frac{\partial T}{\partial t} = div(a_t \nabla T + \delta_t \nabla u + \alpha_t \nabla p) + h_{lv}\rho_s \chi \frac{\partial \omega}{\partial t} \tag{46}$$

$$h_a \frac{\partial p}{\partial t} = div(\lambda_f \nabla p) + \rho_s \chi \frac{\partial \omega}{\partial t} \tag{47}$$

5. Overview of Driving Potential of Coupled Heat and Moisture Transfer

The literature provided different models of coupled heat and moisture transfers in porous materials. The question of choice of appropriate transfer potentials remains a point of discussion for several researchers in the scientific community. Indeed, whatever the adopted transfer driving force (water content, vapor pressure, chemical potential, temperature, liquid pressure, capillary pressure), the formulation of moisture transfer equations are equivalent [32]. Nevertheless, experimentally this multiple choices of moisture driving potential is not always so obvious. Knowing that the water content is not a state variable in the thermodynamic sense. In particular, it is not continuous at the boundary between two different materials, and has a random spatial distribution in a heterogeneous medium. But the driving potential that is most agreed by the experts in the field of construction is the vapor pressure gradient generally caused by wind, pressurization by ventilation, draft effect, etc.

In fact, the vapor pressure potential partially dominates the water transfer assembly in the porous components. Hence the contribution of other transfer driving forces such as the suction pressure associated with the transport of the liquid phase [33]. Moreover, it has been shown that physically the relative humidity gradient causes diffusion through the materia [31]. It has also been shown that vapor pressure, suction pressure and temperature gradient contribute to water transfer in materials [35]. In addition, the orientation of the flows represented by these motors plays an important role on the behavior of the moisture inside the material: the potential of relative humidity contributes less when the gradients of temperature and pressure of vapor are in competitive position. This remark is not valid for all materials, especially hygroscopic insulators [35]. Also, it was quantified the phenomenon of thermodiffusion in building materials, by the identification of a new experimental parameter called thermogradient coefficient [36]. It has been shown that the latter is sensitive in the case of low gradients of vapor content. Recently, Trabelsi has shown the sensitivity of the thermodiffusion phenomenon through an experimental study on building materials [37].

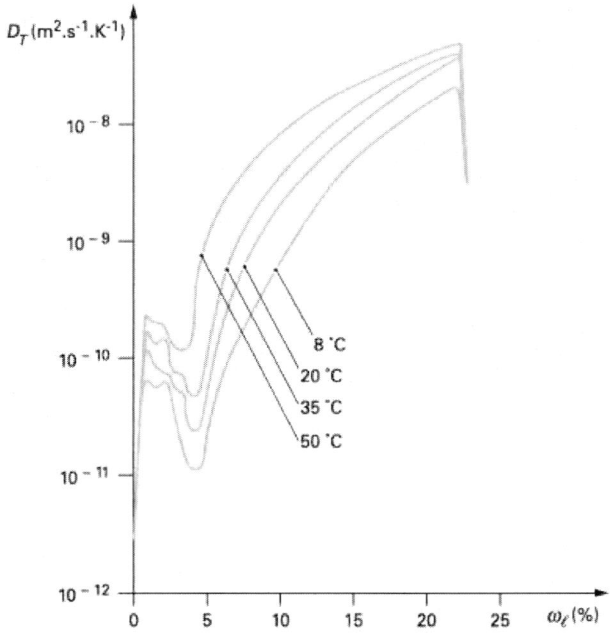

Figure 6. Non-isothermal mass diffusion coefficient versus moisture content of porous materials at four different temperatures [38].

To verify this, the use of other domains, outside the civil engineering, where porous media are exposed to temperature gradients, seems paramount. The studies available on porous media in general shows that the effect of temperature on water transfer is present in porous media [38], [39]. In this context, several points relating to the reality of moisture transfer (vapor or liquid phase) in porous media are in question and need to be examined in more detail. Mass and heat transfers must be considered simultaneously and not separately. These findings led to the following question:

What exactly happens during the transfer of moisture inside porous media?

The behavior of unsaturated porous media has been examined in the framework of classical theories, within the meaning of Fick and Darcy's law [16], [29], [40]. The transfer equations is defined considering the multiphase heterogeneous medium (vapor, liquid, solid and/or gaseous phase) [29]. There is then a whole set of menisci where the capillary effects must be taken into account to characterize the transfer modes relative to each phase. Indeed, coupled moisture and heat transfer phenomena in partially saturated porous materials are diverse and complex. Kelvin's law constitutes a platform for most of these transfer processes (liquid water flow, evaporation drying, capillary condensation, etc.). On the one hand, the capillary transfer occurs in the liquid phase via the capillary forces of interaction between the fluid phases and the solid matrix. On the other hand, the drying phenomenon is dominated by this capillarity, under a slow evaporation, where the displacement of the vapor liquid interface strongly depends on the complex geometry of the microstructure of the considered porous media. In addition, for low relative humidities, a series of evaporative condensation mechanisms occur, resulting in discontinuity of liquid bridges during mass transfer [16]. So far, the literature does not give any more answers about this separation of liquid-vapor phases inside porous building materials. In this sense, a tough question arises as to the precise physical description of the displacement of the liquid-gas interface at the local scale. Nevertheless, a theory attempting to explain these phenomena has been developed [41]. For materials with high porosity, the authors found a significant decrease in vapor pressure. This drop becomes less noticeable in pores of less than 0.1 mm. Also, during the evaporation phase, an excess of enthalpy due to the molar polarization of the liquid and the porous solid is observed. This favors the creation of liquid bridges for fairly high relative humidities [41].

6. Modeling Results

Moisture content and temperature profile evolution within the wall, for both models, are indicated in Figure 7. Globally, a similar reproducibility of the previous model [42] can be seen. The attended variation is explained by the high air permeability of the material [43] and the strong convective coupling of the present model. This means that convection favors drying inside the wood specimen: the drying kinetic is slightly increased according to the present model.

To have a better understanding of these results, the moisture content and temperature profiles inside the overall wood slab were simulated. The convective terms advects moisture content to the surface of the specimen. Simultaneously, this movement decreases the moisture content distribution (vertical drop) which combines the explanation before the increase of drying kinetic.

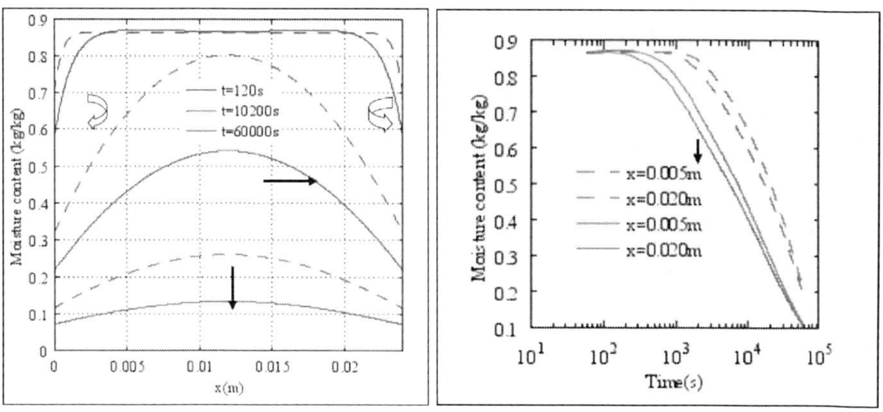

Figure 7. Wood moisture content at different times (a); and different positions (b). (present work is presented in solid line and Remki et al. in dashed line).

Similarly, Figure 8 show the temperature distributions inside the specimen compared for both models. As a first remark, the temperature is slightly affected by the strong coupling of the present model because the mass transfer process is commonly slow than heat transfer. Also, air convection plays a refreshing role because of the trapped air within pores, which explains the temperature decrease in comparison to the model [43].

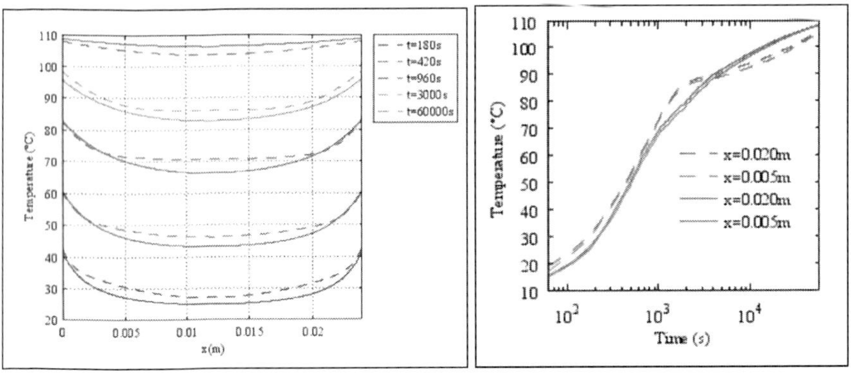

Figure 8. Wood temperature profiles at different times (a); and different positions (b). (present work is presented in solid line and Remki *et al.* in dashed line).

Conclusion

The present study makes it possible to establish a general state of art regarding the hygrothermal transfers in the porous materials of construction. In addition, a brief overview on the reality of the choice of transfer driving potentials has been presented in order to clarify the ideas on the implementation of transfer potentials in mass balance equations. At the end of this chapter, an attempt has been made to explain the combined vapor-liquid transfer in the porous media in order to show the complexity of these phenomena. The most important information from this chapter is the need to optimize transfers in porous building materials with integration of the effect of total pressure. Therefore, the employed diffusive model incorporates the molar transfer of liquid and water vapor phases caused by the total pressure gradient.

References

[1] Bories S. A., Natural convection in porous media, *Adv. Transport Phenom. Porous Media*, 70-144 (1987).

[2] Whitaker S., Simultaneous Heat, Mass, and Momentum Transfer in Porous Media: A Theory of Drying, *Adv. Heat Transfer. Elsevier*, 119-203 (1977).

[3] Keller J. B., Effective behavior of heterogeneous media, *Statistical Mechanics and statistical Method in theory and Application*, U. Landman, ed. New York. Plenum, 631–644 (1977).

[4] Auriault J. L. and Adler P. M., Taylor dispersion in porous media: Analysis by multiple scale expansions, *Adv. Water Resour.* **18**, 217–226 (1995).

[5] Jaffer S. J., Lemaire C., Hansson C. M. and Peemoeller H., MRI: A complementary tool for imaging cement pastes, *Cem. Concr. Res.* **37**, 369–377 (2007).

[6] Millet O., Ait-Mokhtar A. and Amiri O., Determination of the macroscopic chloride diffusivity in cementitious porous materials by coupling periodic homogenization of Nernst-Planck equation with experimental protocol, *Int. J. Multiphys.* **17**, 129–145 (2008).

[7] Bourbatache K., Millet O. and Aït-Mokhtar A., Ionic transfer in charged porous media: Periodic homogenization and parametric study on 2D microstructures, *Int. J. Heat Mass Transfer.* **55**, 5979–5991 (2012).

[8] ISO 13788, Performance hygrothermique des composants et parois de bâtiments- Température superficielle intérieure permettant d'éviter l'humidité superficielle critique et la condensation dans la masse- Méthode de calcul. [Hygrothermal performance of building components and walls - Internal surface temperature to avoid critical superficial moisture and condensation in the mass - Method of calculation.] *In Monolingue*, ICS: 91. 060.01, 91.120.10, 1-34 (2001).

[9] Brunauer S., Emmett P. H. and Teller E., Adsorption of Gases in Multimolecular Layers, *J. Am. Chem. Soc.* **60**, 309–319 (1938).

[10] Langmuir I., The adsorption of gases on plane surfaces of glass, mica and platinum, *J. Am. Chem. Soc.* **40**, 1361 (1918).

[11] Straube J., Moisture and materials, *Building Science Corporation*, BSD-138 906 (2006).

[12] IUPAC, Reporting Physisorption data for Gas/solid Systems, *J. Am. Chem. Soc.* **57**, 369–377 (1985).

[13] Bellini Da Cunha Neto J. A. and Daían J.-F., *Transport d'humidité en matériau poreux en présence d'un gradient de température, Caractérisation expérimentale d'un bâton cellulaire* [*Moisture transport in porous material in the presence of a temperature gradient, Experimental characterization of a cellular cell*], Thése, Université de Grenoble 1, Saint-Martin-d'Höres.

[14] Steeman H. J., Van Belleghem M., Janssens A. and M. De Paepe, Coupled simulation of heat and moisture transport in air and porous materials for the assessment of moisture related damage, *Build. Environ.* **44**, 2176–2184 (2009).

[15] Gertis K., Increased insulation in building - a way to aviod the coming change in climates, *Bauphysiks* **13**, H., S. 132–137 (1991).

[16] Philip J. R. and DeVeries D. A., Moisture movement in porous material under temperature gradients, *Trans. Am. Geophysical Union* **38**, 222–232 (1957).

[17] Kunzel H. M., Simultaneous heat and moisture transport in building components: One- and two-dimensional calculation using simple parameters, *Fraunhofer IRB Verlag Suttgat*, ISBN: 3-8167-4103-7 (1995).

[18] Dutta B. K., Principles of mass transfer and separation processes, *Can. J. Chem. Eng.* **87**, 818–819 (2009).

[19] Bear J., *Dynamics of Fluids in Porous Media*, Dover Publications Inc. (2008).

[20] Dybbs A. and Edwards R. V., A new look at porous media fluid mechanics Darcy to turbulent, *Fundamentals of transport phenomena in porous media Martinus Nishoff*, Dordrecht, 199–256 (1984).

[21] Brinkman H., A calculation of the viscous force exerted by a flowing fluid on a dense swarm of particles, *App. Sci. Res.***1**, 27–34 (1949).

[22] Cheng P., Heat transfer in geothermal systems, *Adv. Heat Transfer.* **14**, 1–105 (1978).

[23] Pedersen C. R., *Heat and Moisture Transfer in Building Constructions*, PhD. Thesis (Report No.214), Thermal Insulation Laboratory, Technical University of Denmark (1990).

[24] Richards L. A., Capilary conduction of liquids through porous mediums, *Physics* **1**, 318–333 (1931).

[25] Crausse P., Laurent J. P. and Perrin B., Influence des phénoménes d'hystérésis sur les propriétés hydriques de matériaux poreux: Comparaison de deux modéles de simulation du comportement thermohydrique de parois de bâtiment [Influence of hysteresis phenomena on the hydric properties of porous materials: Comparison of two simulation models of the thermohydric behavior of building walls], *Revue Générale de Thermique* **35**, 95–106 (1996).

[26] Lewis R. and Ferguson W. J., A partially nonlinear finite element analysis of heat and mass transfer in a capillary-porous body under the influence of a pressure gradient, *App. Math. Model. Ed., rub* **17**, (1993).

[27] Luikov A. V., *Heat and mass transfer in capillary-porous bodies first edition*, Book, Oxford: Pergamon press (1966).

[28] Lewis R. and Ferguson W. J., The effect of temperature and total gas pressure on the moisture content in a capillary porous body, *Int. J. Numer. Methods Eng.* **29**, 3756–369 (1990).

[29] Fengzhi L., Yi L., Yingxi L. and Zhongxuan L., Numerical simulation of coupled heat and mass transfer in hygroscopic porous materials considering the influence of atmospheric pressure, *Fundamentals: Int. J. Comput. Method.* **45**, 249 (2004).

[30] Dos Santos G. H. and Mendes N., Heat, air and moisture transfer through hollow porous blocks, *Int. J. Heat Mass Transfer.* **52**, 2390-2398 (2009).

[31] Abahri K., Bennacer R., Belarbi R., Sensitivity analyses of convective and diffusive driving potentials on combined heat air and mass transfer in hygroscopic materials, *Numer. Heat Transfer., Part A : Applications* **69**, 1079-1091 (2015).

[32] Funk M. and Wakili K. G., Driving Potentials of Heat and Mass Transport in Porous Building Materials: A Comparison Between General Linear, Thermodynamic and Micromechanical Derivation Schemes, *Transport in Porous Media* **72**, 273-294 (2007).

[33] Pedersen C. R., Prediction of moisture transfer in building constructions, *Build. Environ.* **27**, 387-397 (1992).

[34] Padfield T., *role of absorbent materials in moderating changes of relative humidity*, PhD, Thesis, The Technical University of Denmark (1998).

[35] Peuhkuri R., *Moisture Dynamics in Building Envelopes*, PhD, Thesis, Department of Civil Engineering, Technical University of Denmark (2003).

[36] Belarbi R., Qin M., Aït Mokhtar A. and Nilson L. O., Experimental and theoretical investigation of non-isothermal transfer in hygroscopic building materials, *Build. Environ.* **43**, 2154–2162 (2008).

[37] Trabelsi A., Belarbi R., Abahri K., Qin M., Assessment of temperature gradient effects on moisture transfer through thermogradient coefficient, *Build. Simul. Int. J.* **5**, 107–115 (2012).

[38] Mojtabi A., Bories S., Quintard M. and Prat M., Transferts de chaleur dans les milieux poreux-changement de phase [Heat transfer in porous media-phase change], *Techniques de l'ingénieur*, BE8251 (2008).

[39] Davarzani H., Marcoux M. and Quintard M., Theoretical predictions of the effective thermodiffusion coefficients in porous media, *Int. J. Heat Mass Transfer.* **53**, 1514–1528 (2010).

[40] Furmaniak S., Terzyk A. P. and Gauden P. A., The general mechanism of water sorption on foodstuffs - Importance of the multitemperature fitting of data and the hierarchy of models, *J. Food Eng.* **82**, 528–535 (2007).

[41] Abu Al-Rub F. A. and Datta R., Theoretical study of vapor pressure of pure liquids in porous media, *Fluid Phase Equilibria* **147**, 65-83 (1998).

[42] Remki B., Abahri K., Tahlaiti M. and Belarbi R., Hygrothermal transfer in wood drying under the atmospheric pressure gradient, *Int. J. Thermal Sci.* **57(0)**, 135-141 (2012).

[43] Suard S., Lapuerta C., Kaiss A., Porterie B., Sensitivity analysis of a fire field model in the case of a large-scale compartment fire scenario, *Numer. Heat Transfer, Part A: Application* **63 (12)**, 879-905, (2013).

In: Heat Conduction
Editors: J. Hristov and R. Bennacer
ISBN: 978-1-53614-673-8

Chapter 7

ENERGY STORAGE USING PHASE CHANGE MATERIALS

***Xiaoyan Ma*[1,*], *Jianfei Song*[2] and *Bin Liu*[3]**
[1]LMT Cachan, CNRS, ENS Paris-Saclay University,
Président Wilson, Cachan, France
[2]LIMSI-UPR CNRS, Rue John von Neumann,
Orsay, France
[3]Key Lab of Refrigeration Technology,
Tianjin University of Commerce, Tianjin, P. R. China

Abstract

In this chapter the energy storage properties of *Phase Change Materials* (PCMs) will be presented. Energy consumption in buildings has attracted extensive attention due to its huge potential for environmental sustainable development. The key focus is on the efficiency promotion and optimization of energy systems. Among the available methods the application of the PCMs is the most suggested. In this chapter, firstly the basic properties and applications of PCMs will be introduced, which will allow readers be more familiar with the functional role of PCMs. Secondly for the sake of effective analysis during the phase change process, some analytical solutions are described to solve the problems in different cases. Finally in order to obtain and complete the knowledge of phase change mechanism in a bounded domain, a semi-analytical solution is

*Corresponding Author Email: ma@lmt.ens-cachan.fr.

proposed based on the connection method between the short time solution in infinite domain and the the steady-state solution.

Keywords: phase change materials (PCMs), energy storage, bounded domain, semi-analytical solution

1. Phase Change Materials for Energy Storage

The PCMs feature on the phase change process, in this section, the liquid-solid phase change process (either melting or solidification) will be presented. Afterward, the existing analytical solutions for different phase change problems will be illustrated for a better understanding and deeper exploration. In addition, some strategies in simulating the source term of phase change will be introduced.

1.1. General Introduction

In order to improve the energy efficiency and thermal comfort in the buildings, the application of the phase change materials (PCMs) is proposed as one of the suggested solutions [1]. PCMs are able to vary the material's phase state under the different temperatures by absorbing or releasing a large amount of latent heat during the phase change process [2]. Thus the process of changing the material's phase state and varying the physical properties is called the phase change process. Owing to the huge amount of potential heat, once this type of materials are widely used in human life, it will contribute significantly to energy conservation and environmental protection [3], [4], [5]. Nowadays more and more researchers focus on the exploration of the detailed property of PCMs for the purpose of taking full advantage of them.

The liquid-solid phase change process exists widely in many applications and plays important roles in broad fields[6]. Studies dealing with phase change can be found in different situations. However, among these existing solutions there are only a few so-called reference results that are available for the thermal problems [7]. This has attracted increasing interest of numerous investigators. In order to overcome this shortage, researchers commonly validate their results by comparing them with the so-called Stefan analytical solution. It should be noted that this solution is established in the case of one-dimensional phase change in an assumed semi-infinite domain. With the purpose of achieving the completeness of the study and carrying out the analysis of phase change problems in finite domains (a bounded domain or cavity), a semi-analytical solution could be found under the certain additional hypotheses. This solution will be verified by comparing with the numerical results, which will indicate the advantages and limitations.

1.2. Strategies in Simulating the Phase Change Source Term

During the process of solidification, PCMs release heat to the environment, while during the process of fusion (or melting) it will absorb (or store) heat from the environment [8]. The temperature profile versus enthalpy for most PCMs is showed in Figure 1. The energy that can be stored or released when the physical state changes is called phase change heat. During the phase change process, the temperature of the material itself remains almost constant until the phase change process finished [9]. In addition, the latent heat is much higher than the sensible heat, which could be identified in Figure 2. The process of phase change also reflects the relationship between the apparent heat capacity (c_p^{app}) and temperature gradient expressed by the partial differential equation: $\rho c_p^{app} \frac{\partial T}{\partial t} = \alpha \bigtriangledown^2 T$ [10]. The advantage of adopting the apparent heat capacity is to avoid the the complexity in interface discontinuity so that it simplifies the source term. The principle is that the phase change materials store energy during the heat transfer process, so it will prolong the energy transmission time and reduce the temperature gradient.

There are also other approaches to deal with the source term. One of them is to apply the Heaviside function $H_v(T)$ on the interface to illustrate the discontinuity in the enthalpy versus temperature ($\rho c_p \frac{\partial T}{\partial t} = \alpha \bigtriangledown^2 T + H_v(T)$). Another approach is to use a re-meshing on the two sub-domain (domain of liquid or solid) and connect the solid and liquid phase by the interface condition. This method is adopted and illustrated in Section 3.1 by using the equations depicting the solid phase/ liquid phase and the connecting boundary condition on the interface.

2. Properties and Applications of PCMs

In this section, more detailed properties of some typical PCMs will be presented, including the classification, the thermal properties, the advantages and

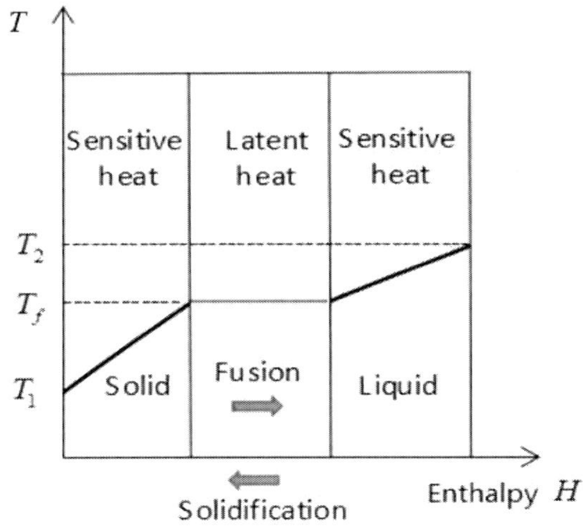

Figure 1. The process of phase change.

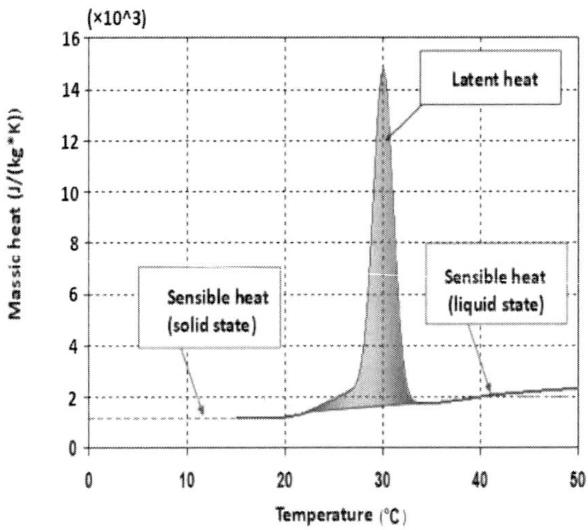

Figure 2. The comparison between latent heat and sensible heat.

shortcomings of each type and some typical applications in buildings and energy storage systems.

2.1. Relative Properties of PCMs

PCMs can be divided into organic PCMs and inorganic PCMs[11], [12], [13]. Organic PCMs include mainly fatty acid and polyol PCMs (Table 1), and also the paraffin PCMs (Table 2). Table 1 shows the properties of different organic fatty acid and polyol PCMs, which indicates that the thermophysical properties of organic PCMs are closely related to the functional groups and the chain length of organic compounds. For the same functional group, the longer the chain length is, the higher the phase transition temperature and the more the latent heat of phase change. Paraffin wax, as another important composition of organic PCMs, is a kind of by-product of refined petroleum. It is usually derived from the wax fraction of crude oil and it needs to be extracted from oil by several processes such as atmospheric and vacuum distillation, solvent refining, solvent dewaxing and deoiling, hydrorefining, etc. [14]. Table 2 lists the relationship among the number of carbon atoms, melting point and the latent heat of melting for paraffin. With the increase of the carbon atoms chain, the melting temperature increases, and also the latent heat globally increases. It's important to have a better knowledge about the property of PCMs before they can be used more effectively. For organic phase change materials, more desirable PCMs can be obtained by synthesizing multiple systems.

Table 1. The properties of fatty acid and polyol PCMs

Materials	Formula	Melting point (^oC)	Latent heat (kJ/kg)
Acetic	CH_3COOH	16.7	184
Polyethylene glycol 600	$H[OC_2H_2]_n - OH$	20-25	146
Capric acid	$CH_3(CH_2)_8 - COOH$	36	152
Elaidic acid	$C_8H_7C_9H_{16} - COOH$	47	218
Lauric acid	$CH_3(CH_2)_{10} - COOH$	49	178
Pentadecanoic acid	$CH_3(CH_2)_{13} - COOH$	52.5	178
Tristearin	$(C_{17}H_{35}COO)C_3H_5$	56	191
Myristic acid	$CH_3(CH_2)_{12} - COOH$	58	199
Plamitic acid	$CH_3(CH_2)_{14} - COOH$	55	163
Methyl fumarate	$(CHCO_2CNH_3)$	102	242

Table 2. The properties of paraffin PCMs

Number of carbon atoms	Melting point (^oC)	Latent heat of melting (kJ/kg)
14	5.5	228
16	16.7	237.1
18	28	244
20	36.7	246
22	44	249
24	50.6	255
26	56.3	256
28	61.6	253
30	65.4	251
32	69.5	170
34	75.9	169

Inorganic PCMs, as another main constitute of PCMs, are characterized by higher energy storage density, smaller volume change during the phase change process, and lower cost. The contents include mainly alkali, alkaline earth metal halides and hydrates of salts such as sulfates, phosphates, nitrates, acetates, and carbonates, as showed in Table 3. However, this type of material is prone to be over-cooled and phase separation appears in the phase transition process. So it is necessary to add anti-supercooling agent and anti-phase-separator to enhance its stability and prolong its service life. For example, the addition of *NaCl* and excessive water into $CaCl_2 \cdot 6H_2O$ can keep $CaCl_2 \cdot 6H_2O$ in good stability. The properties of some commonly used inorganic PCMs are listed in Table 4. The specific heat represents for the required heat energy to increase 1 ^oC for the substance quantity of 1 kg. Different substance is characterized by different specific heat capacity which is the physical property of a matter. This value can be obtained by dividing volumetric heat capacity with the density. Table 4 suggests also that water has the highest value of the specific heat among these PCMs, which means under the same cooling or heating condition, the temperature variation for water is the smallest. Owing to this advantage, water is widely used in the daily life of human being, such as the cooling of the car's engine and hot water heating commonly used in winter.

Compared with the inorganic PCMs, the organic PCMs is absent of over-cooling or precipitating phenomena, and with relatively stable performance. However, there are usually some shortcomings like low thermal conductivity, low density, and poor thermal storage capacity per unit volume [15].

Table 3. The properties of hydrated salts PCMs

Saline hydrates	Melting point (^oC)	Latent heat of melting (kJ/kg)
Glauber salt ($Na_2SO_4 \cdot 10H_2O$)	32.4	252
Calcium chloride, hexahydrate $CaCl_2 \cdot 6H_2O$	27-30	190
Sodium acetate trihydrate $NaCH_3COOH \cdot 3H_2O$	58	170-264
Barium hydroxide octahydrate $Ba(OH)_2 \cdot 8H_2O$	78	301
$NH_4 \cdot Al(SO_4)_2 \cdot 12H_2O$	95	238.5
Sodium phosphate dodecahydrate $Na_3 \cdot PO_4 \cdot 12H_2O$	77	213
Sodium pyrophosphate decahydrate in capsule form (Calotherm 70)	70	184
Sodium carbonate decahydrate $Ca(NO_3)_2 \cdot 4H_2O$	40-43	140

Table 4. The properties of commonly used inorganic PCMs

Materials	Density (kg/m^3)	Specific heat ($J/(kg \cdot K)$)	Volumetric heat capacity ($J/(m^3 \cdot K)) \times 10^6$
Clay	1458	879	1.28
Bricks	1800	837	1.51
Sandstone	2200	712	1.57
Wood	700	2390	1.67
Concrete	2000	880	1.76
Glass	2710	837	2.27
Aluminum	2710	896	2.43
Iron	7900	452	3.57
Steel	7840	465	3.68
Water	988	4182	4.17

Several assumptions of the flow are made to simplify and analyze the phase change process, including laminar and two-dimensional flow pattern, Newtonian and incompressible liqid and constant physical properties of the flow except density in the buoyancy term. A typical mode of the phase change interface is showed in Figure 3. It indicates that the flow in liquid phase could induce irregular shape on the interface, and such coupling among conduction, convection and phase change adds complexity in predicting the energy stored or released over time. In order to avoid such flow, a popular way is to mix with additive to generate PCMs gel. The practical mixture has the same characteristic as that used in the previous research work of Royon et al. [16]

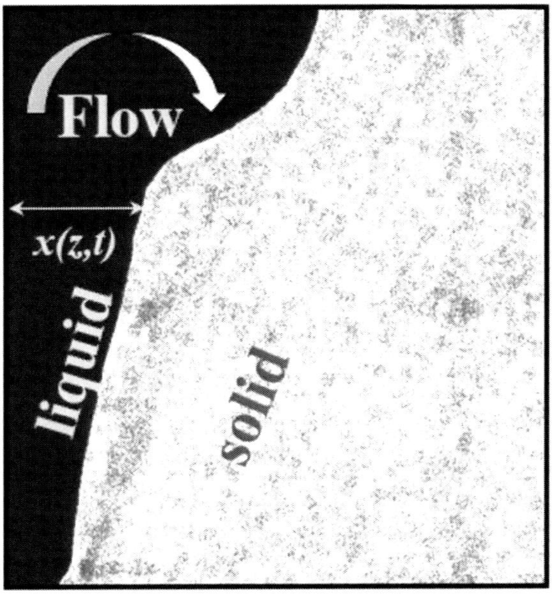

Figure 3. A typical mode of phase change interface.

2.2. Applications of PCMs

The incorporation technique of PCMs into the wall of buildings is subject of numerous works and widely studied[17]. However, the anti-symmetric character of storing/releasing energy management in walls is less controlled and most of the time does not fit with the designed optimal conditions. A solution is proposed in this study to overcome the anti-symmetry problem and to fit with different ambient conditions. It is based on the direct integration of the stabilized PCMs (gel) in a building material including ventilation tubes. Such gel could prevent the unexpected flow in liquid domain and eventually avoid any possible leakage of the materials. The objective is the optimization of a wall combining the heavy inertia induced by the PCMs, intra-ventilation control and the contribution to air renewal.

The developments are very fast for energy and thermal system, and also for the innovation of construction materials . Among all the challenges of energy consumption, the thermal buildings are one of the biggest consumers of energy [18]. Efficiency and optimization of the system remain the main focus in decreasing energy consumption and improving system function of the related research work. Balancing accessible energy with inhabitant comfort requires more innovative and effective technological solutions [19]. In the housing sector, cost reduce and space optimization are the two key reasons for reducing the thickness of the wall in new buildings. However, the reduction of the wall thickness significantly affects the thermal inertia of the frame and makes it inadequate to suppress the oscillation, which is caused by variation of outdoor temperature. In addition, the thermal comfort management through the control system acting on ventilation could induce higher power consumption and the disadvantages associated with the operation of these devices (noise, maintenance, etc.). A typical phase change material application in ventilated wall energy storage is showed in Figure 4. By increasing the thermal inertia, PCMs regulate the ambient temperature and store/release energy in cyclic way.

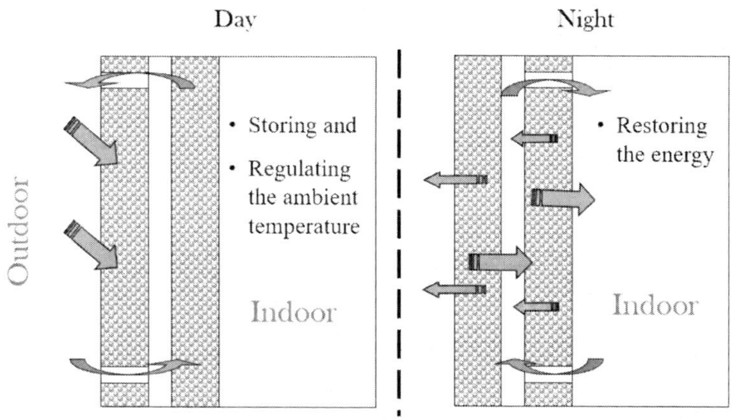

Figure 4. A schematic of PCMs applied in ventilated wall energy storage.

The incorporation of PCMs in building envelopes is acting as a motivating way to improve energy feedings [20]. When joined into construction elements, latent heat becomes an important index to reflect the energy storage capacity. The thermal phase shift happens during the process of energy storage into the damped recovery phase, which diminishes the request based on the outdoor climatic situations of the building under consideration. Numerous studies on this subject have been done aiming at answering all kinds of difficulties that are against the optimal use of this solution [21], [22]. About these limitations, one can choose the appropriate way to merge these products. Direct integration, impregnation of building materials, and incorporation of capsules containing PCMs are among the popular methods. Hence, in order to reduce the threat of exposure during the melting process of these products, we use mixtures which are designed to strengthen the PCMs matrix, such as the use of polymers, a possible way to have a mechanically stable gel during the process of liquid phase [15]. Likewise, The high efficiency of the developed system demands an optimized study of the type of PCMs and the size of the system and probable mixtures, so that we can appropriately achieve the energy charge/discharge in these products [23].

Two types of integration are distinguished when it is combined into building envelopes or in applications which are associated to thermal comfort adjustments during the occupancy of a building. In the first type of integration, the following cases are pointed out: integration in floors [24], walls [25], masonry elements [26] and roofs [27], as well as in windows and glazing [28]. With regard to the use of PCMs in comfort adjustment systems within the building, the free-cooling method need to be paied more attention, which is based on heat exchangers developed according to a couple of methods [29], [30]. Based on these studies, we can conclude that the involvement of PCMs increases thermal inertia and improves the air quality of buildings.

The studies of the vertical members of the building with PCMs occupy the majority, and it is notable that the wall panels can be linked to the base walls. In this situation, many products have been examined and sold on the market because of the advantage of diminishing the level of the peak temperature during hot periods and saving energy. The damping of the peak temperature achieved could rise to 4^oC [31]. PCM-based concrete and coatings are part of the solutions applied to the wall. They are established under several compositions with different percentages of PCMs. The tests conducted in a real scale on cells show the functional relationship between the storage capacity and the volume ratio of the PCMs, and also the temperature difference in indoor surface with respect to a reference cell [32].

3. Some Solutions in Phase Change Problems

The analytical solutions exist in different situations and much of them are validated by the numerical method. In this section, firstly the analytical solution in the semi-infinite domain will be recalled, which is well known as 1D Stefan problem. The short time resolution of two-dimensional corner problem during phase change will be presented afterwards. The corner problem corresponds in several situations to the interface front position evolution over the square domain, which is also investigated by numerical method. Another approach in such square domain consist of an analysis of the system and estimation of the energy extracted (injected) during the phase change process. After that, a more global analytical solution of energy extracted during freezing process is mentioned based on the discussion of different *Biot* numbers. In addition, the other case occurred in bounded domain with different temperatures on both side, will be illustrated afterwards. A semi-analytical solution is proposed, which is mainly a combination of two basic 1D problem with a connecting expression along the diagonal distance from the corner. At last, the solution in multi-phase of Cartesian domain and analysis in cylindrical case will be presented.

3.1. Analytical Solutions in Semi-Infinite Domain

The classical Stefan problem is a solidification-melting problem, such as transformation between ice and water. Due to the evolving interface is a priori unknown, the determination the interface frontier will be a part of the solutions.

A one-dimensional Stefan problem in semi-infinite domain is illustrated in Figure 5. The imposed boundary conditions are constant temperature T_1 on one side, and initial temperature $T_{initial}$ is uniform in all the domain. Function $X(t)$ denotes the position of the solid-liquid interface frontier. Thus the advancing direction of the interface with a constant temperature T_1 will rely on the condition either it is solidification or melting process.

In the following discussion, the thermal properties of heat capacity, thermal conductivity, and thermal diffusivity are denoted by c_1, k_1, α_1 in solid phase and c_2, k_2, α_2 in liquid phase respectively. In addition, L_H, T_f and ρ represent for the latent heat, the melting temperature(fusion temperature) and the density of the substance (assumed to be the same in both phases), respectively.

Therefore the classic one-dimensional Stefan problem of stationary case can be described by the following equations.

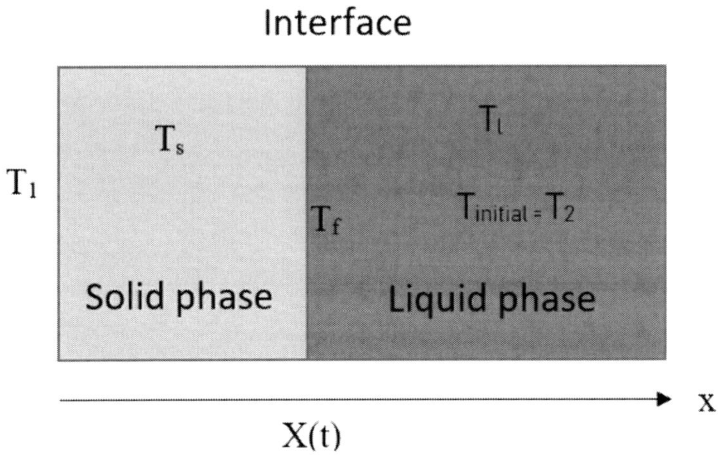

Figure 5. 1D Stefan problem in semi-infinite domain.

$$\frac{\partial^2 T_s}{\partial x^2} - \frac{1}{\alpha_1}\frac{\partial T_s}{\partial t} = 0,\ x < X(t) \tag{1}$$

$$\frac{\partial^2 T_l}{\partial x^2} - \frac{1}{\alpha_2}\frac{\partial T_l}{\partial t} = 0,\ x > X(t) \tag{2}$$

With the boundary conditions: $T_s = T_1$, when $x = 0$, $T_l = T_{initial} = T_2$, when $x \to \infty$.

At the interface, there is: $T_s(t, x) = T_l(t, x) = T_f$ it also follows:

$$L_H \cdot \rho \frac{dx}{dt} = k_1 \frac{\partial T_s}{\partial x} - k_2 \frac{\partial T_l}{\partial x} \tag{3}$$

And the initial conditions are:

$$T_l(t = 0, x) = T_2, X(t = 0) = 0 \tag{4}$$

The analytical solution of these formulas has been proved very complicated due to the coupling and the non-linear character, and it still remains unknown except in the case of semi-infinite domain. Now the problem is simplified to one-dimensional problem associated with the two temperatures T_1 and T_2. Thus the frontier of the asymptotic value is governed by the equation of the one-dimensional Stefan problem [33].

$$X(t) = 2\lambda\sqrt{\alpha_1 t} \tag{5}$$

In which, λ is the root of the following equation:

$$\frac{e^{-\lambda^2}}{erf(\lambda)} - (\frac{\alpha_2}{\alpha_1})^{1/2} \cdot \frac{k_2}{k_1} \cdot \frac{T_2 - T_f}{T_1 - T_f} \cdot \frac{e^{-\lambda^2 \cdot (\frac{\alpha_2}{\alpha_1})^{1/2}}}{erfc(\lambda \cdot (\frac{\alpha_2}{\alpha_1})^{1/2})} = \sqrt{\pi}\frac{L_H \cdot \lambda}{c_1(T_f - T_1)} = \sqrt{\pi}\frac{\lambda}{Ste} \tag{6}$$

Where $Ste = \dfrac{c_1(T_f - T_1)}{L_H}$ represents for the Stefan number.

Taking a solidification case as an example, an analytical solution of the interface evolution under different Stefan numbers and different parameter ratios in semi-infinite domain is presented in Figure 6.

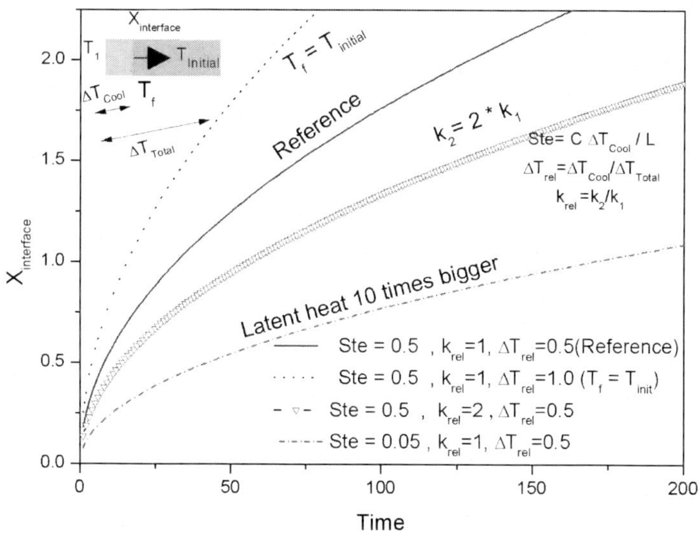

Figure 6. Interface evolution of solidification case in semi-infinite domain.

The interface evolution indicates that in comparison with the reference case, if the ΔT_{rel} is increased, which is the ratio between the cooling temperature difference ΔT_{cool} ($\Delta T_{cool} = T_f - T_1$) and total temperature difference ΔT_{total} ($\Delta T_{total} = T_2 - T_1$), the advancing of the interface will be faster, which is reasonable due to the higher temperature gradient of the freezing process. The blue line shows the results when the thermal conductivity ratio between the liquid and solid phase is doubled, the movement of the interface frontier will be slowed down. In order to analyze the latent heat effect, a red dashed line is drawn by reducing Stefan number 10 times smaller, which means the latent heat value takes dominant role. It is obvious that the process of solidification is much slower than the reference one. This kind of PCMs would have more advantages in applications which requires more stable temperature fluctuation.

3.2. Short Time Solution in Rectangular Corner

Likewise in the one-dimensional case, the analytical solution also exists on some conditions of two-dimensional problem. This conventional problem is related to the solidification of a semi-infinite rectangular corner ($x > 0, y > 0$), which is initially liquid at uniform temperature $T_i(T_i > T_f)$ since the beginning ($t = 0$) of the solidification process, a constant temperature T_0 ($T_0 < T_f$) is imposed on both sides of the corner. The imposed temperature on each side could also be different as T_1 and T_2, as showed in Figure 7 and Figure 8. It's the solidification process of a rectangular corner.

The analytical solution of this problem is characterized by the following dimensionless parameters:

$$X = \frac{x(t)}{2\sqrt{\alpha_1 t}}, Y = \frac{y(t)}{2\sqrt{\alpha_1 t}} \tag{7}$$

The solution properties have been studies by Budhia, Kreith [34] and Rathjen, Jiji [35]. They pointed out that the position of the front following the diagonal is in the form of the Neumann solution.

$$x(t) = y(t) = 2\lambda\sqrt{\alpha_1 t} \tag{8}$$

When the position of the front is in parallel with one of the two walls, the solution is very close to the Neumann solution in the unidirectional case in a short time, i.e., for a fixed x, we have $X = \frac{x(t)}{2\sqrt{\alpha_1 t}}$, $Y = \frac{y(t)}{2\sqrt{\alpha_1 t}}$. For the t in a weak form where λ can be given by the Neumann solution. The complete resolution of the wedge problem leads to analytical formulas that are not

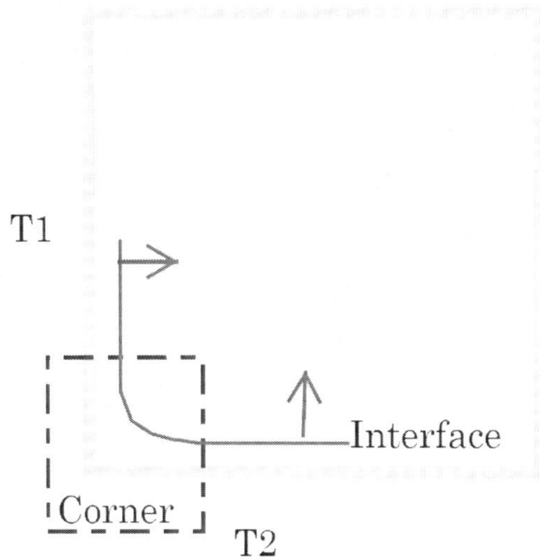

Figure 7. Schematic of body solidification problem.

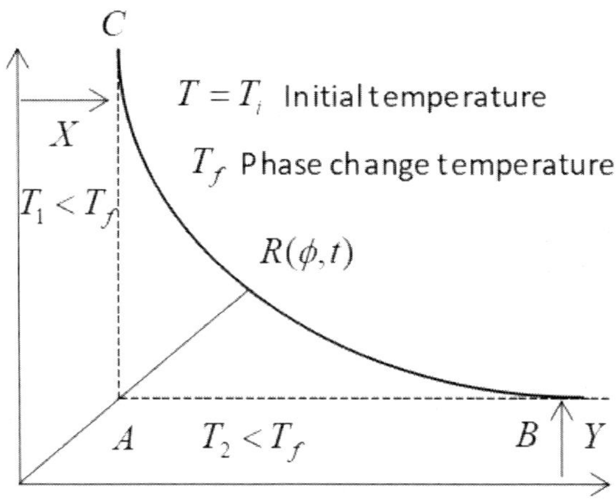

Figure 8. Interface profile on the corner.

directly calculable. Indeed, the distribution of the temperature can be written in the form of a Green integral that must be evaluated numerically in the particular case, in which the thermal diffusivities are supposed to be equal [36]. Nevertheless, the shape of the front can be assumed to be hyperbolic. This hyperbola is asymptotic to both straight lines AC and AB. It is necessary to determine these two straight lines. It worth noticing that on the position that far from the straight line AC, the influence of T_2 is negligible. The problem is reduced to the problem of one-dimensional Stefan associated with the two temperatures T_1 and T_2. Thus the advance of the asymptote AC is governed by the equation of the one-dimensional Stefan problem, which is the same as the semi-infinite domain solution.

It is also notable that the evolutions $X(t)$ and $Y(t)$ are identical because of our choice of boundary conditions. We notice that $R(\phi, t)$ is a point of the front. By using the method of similar solutions [36], we can get the $R(\phi, t)$ in the following form:

$$R(\phi, t) = 2\sqrt{\lambda_1 t} R_0(\phi) \tag{9}$$

Where $R_0(\phi)$ is the position angle that is a function depending only on ϕ. The hyperbole describing the front is of the type:

$$(x - X_1)(y - Y_1) = a(*) \tag{10}$$

With $x = Rcos\phi = 2\sqrt{\lambda_1 t} R_0(\phi) cos\phi$ and $y = Rsin\phi = 2\sqrt{\lambda_1 t} R_0(\phi) sin\phi$.

So the equation of this hyperbola becomes:

$$(R_0(\phi)cos\phi - \lambda_1) \cdot (R_0(\phi)sin\phi - \lambda_2) = \frac{a(*)}{2\sqrt{\alpha_1 t}} \tag{11}$$

What worth mentioned is that the obtained solution is somehow limited when the front line is getting close to the core of the domain. Because limited to the 1D approach, the change of the inner domain is subjected to volume change and so inducing an accelerating phase change due to the curvature increase.

3.3. Energy Analysis in Global Freezing Process

Some explicit solution are proved exist and energy analysis has been studied for some specific cases in a more global boundary condition. For instance, when the substance can be considered as lumped body or isotherm, the solution can be obtained during the freezing process. In Bart's work [33], [37], it is suggested that the dimensionless numbers Ste, Bi and Fo fit with the following equation:

$$\frac{Q_r^2}{2Ste} + \frac{Q_r}{BiSte} - Fo = 0 \tag{12}$$

An interpolation method is applied in Bart's study. Assigning $Q_r = 1$ will yields the well-known Plank equation for a slab.
For $Bi \to 0$ equation 12 transfers into:

$$(BiFo)_{Bi \to 0} = \frac{Q_r}{Ste} \tag{13}$$

and for $Bi \to \infty$:

$$(Fo)_{Bi \to \infty} = \frac{Q_r^2}{2Ste} \tag{14}$$

The equation can be written as:

$$Fo = \frac{(BiFo)_{Bi \to 0}}{Bi} + (Fo)_{Bi \to \infty} \tag{15}$$

For the interpolation equation, a general property is that the exact results are obtained at the limits $Bi \to 0$ and $Bi \to \infty$, and between the two limits it behaves smoothly.

Here Q_r is the ratio of the evacuated heat power during the freezing to the total heat power needed to be extracted (W/m^3), named as relative extracted heat, Bi is the Biot number, Ste is the Stefan number and Fo is the Fourier number. The expressions are given respectively below:

$$Bi = \frac{\alpha(V/S)}{k} \tag{16}$$

$$Ste = \frac{C_p(T_s - T_x)}{L_H} \tag{17}$$

$$Fo = \frac{kt}{\rho c_p (V/S)^2} \tag{18}$$

Where ρ is the density (kg/m^3), C_p is the specific heat capacity ($J/(kg \cdot K)$), and k is the thermal conductivity ($W/(m \cdot K)$), V is the volume (m^3), S is the surface area (m^2), L_H is the latent heat (J/kg), T_s denotes the solidification temperature and T_x is the final temperature. Therefore the solution of Q_r can be presented below:

$$Q_r = \frac{-\frac{1}{B_i Ste} + \sqrt{(\frac{1}{B_i Ste})^2 + 2 \times \frac{1}{Ste} \times F_o}}{\frac{1}{Ste}} \tag{19}$$

Because the removed sensible heat is neglected in comparison to the evacuated latent heat during the phase change, so Q_r can be calculated as follows:

$$Q_r = 1 - (1 - \frac{2b}{L})^2 \tag{20}$$

Where L is the length of the side (m), b is the distance between the solid-liquid interface and the boundary (m). From the equation above, b can be calculated as the following expression:

$$b/L = \frac{1 - \sqrt{1 - Q_r}}{2} \tag{21}$$

The relative extracted heat Q_r versus dimensionless time Fo between initial state($Q_r = 0$) and the end state($Q_r = 1$) is presented in Figure 9.

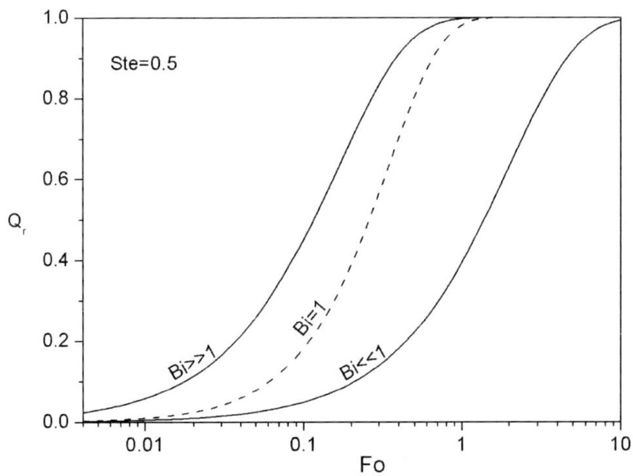

Figure 9. Relative extracted heat Q_r evolution for different Bi numbers.

3.4. A Semi-Analytical Solution in Bounded Domain

In this part a semi-analytical approach for the phase change problem in a bounded domain will be illustrated. The objective is to predict the evolution of the interface throughout its solidification or melting process for the time range of $t \in (0, +\infty)$ [38]. The methodology will be presented in detail in a Cartesian domain. This approach is based on a connection method between solutions at the first moment(short time solution), intermediate solution and the asymptotic value. The short time solution is comparable to the solution in a semi-infinite domain, where the thermal turbulence from the other boundary is negligible, and asymptotic values can be obtained by analysis of the steady state (stationary case) in a long time solution. Furthermore, the semi-analytical result obtained will be compared with the numerical results in order to see the validity limits of the hypotheses which will make it possible to obtain the semi-analytical solution.

3.4.1. Calculation of Short Time Solution

For the validation of the numerical results (obtained in the bounded case), we are more interested in the evolution in the short time period where the interface evolves as if the domain is semi-infinite. We recall that when solving Stefan's problem, it is considered as a fact that when $x \to \infty$, the term $erf(\frac{x}{2\sqrt{\alpha_2 t}})$ tends to be 1. The domain can therefore be considered as semi-infinite if $\frac{L}{2\sqrt{\alpha_2 t}}$ is very large, which also needs to follow the validity condition for the hypothesis of a semi-infinite domain in terms of time. Thus the relationship can be expressed as $t \ll \frac{L^2}{4\alpha_2}$.

3.4.2. Calculation of the Interface Position in the Stationary Case

In the bounded case we assume that there's symmetrical boundary conditions on both sides, then this problem could adapt to a permanent solution, which will be presented below [39]. The temperatures T_1 and T_2 are invariant with respect to y, so in the case of a diffusive transfer we can obtain the temperature field only as a function of x, and the partial derivatives with respect to y and t are zero $(\frac{\partial}{\partial y} = 0, \frac{\partial}{\partial t} = 0)$.

This situation of permanent solution in a unidirectional Cartesian configuration corresponds to the case of the linear temperature profile. A classical situation of pure diffusion is represented by Figure 10.

On the solid side we have: $T_s(x) = a_1 x + b_1$.

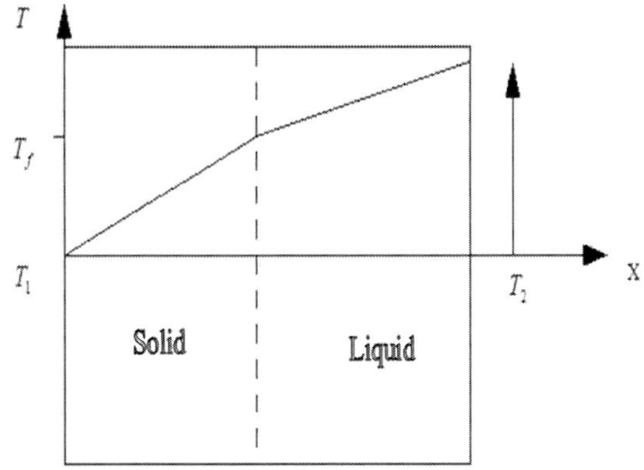

Figure 10. Domain representation and the steady-state temperature profile.

On the liquid side we have: $T_l(x) = a_2 x + b_2$.

The imposed boundary conditions are: $T_s(x = 0) = T_1, T_l(x = L) = T_2$.

The condition at the interface follows: $T_s(x) = T_l(x) = T_f$.

On the other hand, the balance at the interface is written in the stationary mode:

$$k_1 \frac{\partial T_s}{\partial x} - k_2 \frac{\partial T_l}{\partial x} = L_H \cdot \rho_1 \frac{dX}{dt} = 0 \tag{22}$$

$$k_1 \frac{T_f - T_1}{X} - k_2 \frac{T_f - T_2}{X - L} = 0 \tag{23}$$

From these boundary conditions and the interface balance equations we obtain the spatial positions of the interface in steady state. This position will be called asymptotic position, because achieving this value rigorously requires theoretically infinite time. Of course, in practice, it is generally considered that this position is reached when the relative difference is less than a set-point of the order of 10^{-2}. The asymptotic value is therefore given by:

$$X = \frac{L}{1 - \dfrac{T_f - T_2}{T_f - T_1} \cdot \dfrac{k_2}{k_1}} \tag{24}$$

The condition $X < L$ remains satisfied, since: $T_2 > T_f$, and $T_f > T_1$.

The temperature profiles of the solid and liquid sides are given respectively by:

$$T_s = [(T_f - T_1) - (T_f - T_2)\frac{k_2}{k_1}] \cdot \frac{X}{L} + T_1 \tag{25}$$

$$T_l = [(T_f - T_1)\frac{k_1}{k_2} - (T_f - T_2)](\frac{X}{L} - 1) + T_2 \tag{26}$$

Thus, the position of the interface and the temperature distributions in the two phases are established in the case of the stationary regime. These two extreme situations presented above in the bounded domain allow to carry out an analysis both at the beginning and at the end of the evolution process in time.

3.4.3. Stefan's Problem in a Rectangular Enclosure

In this part, we consider a rectangular enclosure $[0, L] \times [0, Y_0]$ that contains a fluid with initial temperature of T_2 ($T_2 > T_f$), as illustrated in Figure 11 [40]. The two horizontal walls are adiabatic. One imposes brutally a temperature T_1 on the left wall ($x = 0$) at the initial time $t = 0$, and the temperature T_2 remains constant on the right wall ($x = L$). For $t \geq 0$, these conditions will induce the appearance of a process of solidification which results in the advance of a solid front in one direction of the absence of flow.

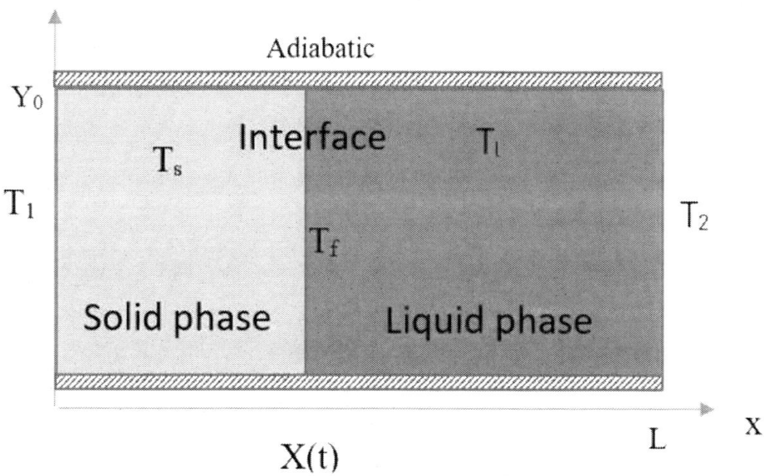

Figure 11. Stefan's problem in a rectangular enclosure.

This is the classic one-dimensional Stefan problem already discussed in the above stationary case which will follow the relationship below.

$$\frac{\partial^2 T_s}{\partial x^2} - \frac{1}{\alpha_1}\frac{\partial T_s}{\partial t} = 0,\ x < X(t) \tag{27}$$

$$\frac{\partial^2 T_l}{\partial x^2} - \frac{1}{\alpha_2}\frac{\partial T_l}{\partial t} = 0,\ x > X(t) \tag{28}$$

With the boundary conditions: $T_s = T_1$, when $x = 0$; $T_l = T_2$, when $x = L$.

And at the interface, where $T_s(t,x) = T_l(t,x) = T_f$, it also follows:

$$L_H \cdot \rho \frac{dx}{dt} = k_1 \frac{\partial T_s}{\partial x} - k_2 \frac{\partial T_l}{\partial x} \tag{29}$$

While the initial conditions are:

$$T_l(t=0,x) = T_2, X(t=0) = 0 \tag{30}$$

The analytical solutions of these equations have been proved to be very complicated due to the coupling and the non-linear character, and it still remains unknown except in the semi-infinite case [41]. However, if we distinguish between the initial time and the later time, we can provide a simplified solution for t ranging from 0 to $+\infty$ by using an adequate approximation in the intermediate time.

First, for the initial time t ($t \in [0, t_0]$), (t_0 will be calculated later), the effect of the right surface ($x = L$) on the advance speed of the front is considered negligible. therefore the position of the front can be obtained approximatly by the solutions corresponding to the solidification problem in a semi-infinite region, which has been explained in the previous section. Obviously this solution, often used in literature to validate numerical models [38], is valid sufficiently only for a short period of time. That is to say, when the front remains far enough from the right wall and does not being affected by the thermal effect caused by the right wall, the evolution of the front is similar to that of the semi-infinite domain.

Secondly, we recall that when t tends to infinity, $X(t)$ tends to $+\infty$, and in the bounded domain the solidification front will tend to an asymptotic position since the right wall is maintained at a fixed temperature. Thus when t tends to

$+\infty$, the asymptotic value of $X(t)$ can be given in the stationary case by:

$$\lim_{x \to +\infty} X(t) = X_{\infty} = \frac{L}{1 + \dfrac{k_2}{k_1}\dfrac{T_2 - T_f}{T_f - T_1}} \tag{31}$$

Eventually, we consider the intermediate time t, which is between the short time and the time that the front reaches the asymptotic value. The desired solution will be connected to the solution obtained at the first moment. To determine this new solution, we mainly base ourselves on the following hypothesis: with time increasing, the solidification front moves with a lower and lower speed. In other words, if one places oneself within the framework of a quasi-stationary approximation, the temperature distribution in the solid and the liquid corresponds to that of a pseudo stationary state. This hypothesis can be expressed physically by the condition that the speed of the energy diffusion is superior to the speed of the front advance. Under this assumption the temperatures in pseudo steady state are given by:

For $x \geq X(t)$:

$$T_s(x,t) = T_1 + \frac{T_f - T_1}{X(t)}x,\ x \geqslant X(t) \tag{32}$$

For $x \leq X(t)$:

$$T_l(x,t) = T_2 + \frac{T_f - T_2}{X(t) - L}(x - L),\ x \leqslant X(t) \tag{33}$$

Which could also be rewrited as the following equality:

$$k_1\frac{T_f - T_1}{X(t)} - k_2\frac{T_f - T_2}{X(t) - L} = L_H \cdot \rho \cdot \frac{dX(t)}{dt} \tag{34}$$

Therefore, the position of the interface $X(t)$ is obtained by solving the following differential equation:

$$\frac{dX(t)}{dt} = \beta_0 \frac{X(t) - X_{\infty}}{X(t)(X(t) - L)},\ t \geq t_0 \tag{35}$$

$$X(t_0) = 2\lambda\sqrt{\alpha_1 t_0},\ t = t_0 \tag{36}$$

$$\beta_0 = \frac{K_1 L(T_f - T_1)}{\rho \cdot L_H \cdot X_\infty} \tag{37}$$

This differential equation of the first order is easy to solve numerically. The Euler scheme can be used for its resolution:

$$\frac{X(t_{n+1}) - X(t_n)}{\triangle t} = \beta_0 \frac{X(t_n) - X_\infty}{X(t_n)(X(t_n) - L)} \tag{38}$$

Now it remains to determine t_0: the connection time between the model of the beginning and the quasi stationary model. This time is determined to ensure the continuity of the front position and its speed.

$$(\frac{dx}{dt})_{t_0^-} = (\frac{dx}{dt})_{t_0^+} \tag{39}$$

so

$$\frac{\lambda\sqrt{\alpha_1}}{\sqrt{t_0}} = \beta_0 \frac{X(t_0) - X_\infty}{X(t_0)(X(t_0) - L)},\ X(t_0) = 2\lambda\sqrt{\alpha_1 t_0} \tag{40}$$

and it results in

$$t_0 = \frac{1}{4\lambda^2\alpha_1}(L - \frac{\beta_0}{\beta_0 - 2\lambda^2\alpha_1}(L - X_\infty))^2 \tag{41}$$

To identify the connection time t_0, we assumed the equality of the velocities between the case obtained under the quasi-stationary hypothesis and the semi-infinite case. These two situations presenting the same speed have differences in the temperature fields, which will induce an incorrect evolution during the transition between the two temperature fields. The effect of this difference should be totally negligible as the hypothesis we have pointed out is true. This passage is intended to reach the accuracy of the solution. In conclusion, the position of the front can is given by:

$$X(t) = 2\lambda\sqrt{\alpha_1 t},\ t \in [0, t_0] \tag{42}$$

$$\frac{dX(t)}{dt} = \beta_0 \frac{X(t) - X_\infty}{X(t)(X(t) - L)},\ t \in [t_0, +\infty] \tag{43}$$

The corresponding solutions in the two particular situation will be the breakthrough point and promote us to manage an analysis in the finite domain case. The results of interface position $X_{interface}$ profile for the three stages are presented in Figure 12, and a comparison between the analytic solution and the numerical solution about the front position versus time is showed in Figure 13. It indicates the favorable fitting between the semi-analytical validation and the numerical results.

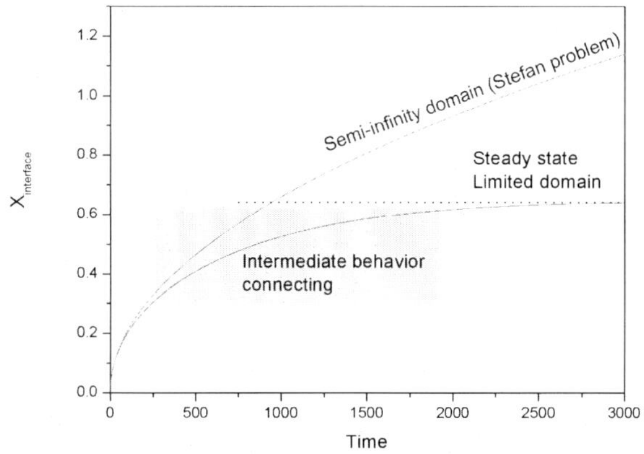

Figure 12. A combination solution for bounded domain.

3.5. Multi-Phase Problem in a Cartesian Domain

Under some circumstances, it is assumed that the material has two transition temperatures T_{f_1} and T_{f_2}, at which the latent heats L_{H_1} and L_{H_2} will be released at each of their interfaces [22]. It is under the consideration that exactly the same method is applied as the previously described: $T_1 < T_{f_1} < T_{f_2} < T_2$. The phases are denoted by solid, mixture and liquid respectively: (T_1, T_{f_1}), (T_{f_1}, T_{f_2}) and (T_{f_2}, T_2).

Using the same analysis as in the single-phase change problem, the following expressions of the two front positions can be found, they can be denoted by $X_1(t)$ and $X_2(t)$, respectively:

$$X_1(t) = 2\lambda_1\sqrt{\alpha_1 t}, \; for\; t \in [0, t_1]$$
$$\frac{dX_1(t)}{dt} = \frac{\beta_1}{X_1(t)} + \frac{\beta_2}{X_1(t) - X_2(t)}, \; for\; t \in [t_1, +\infty] \tag{44}$$

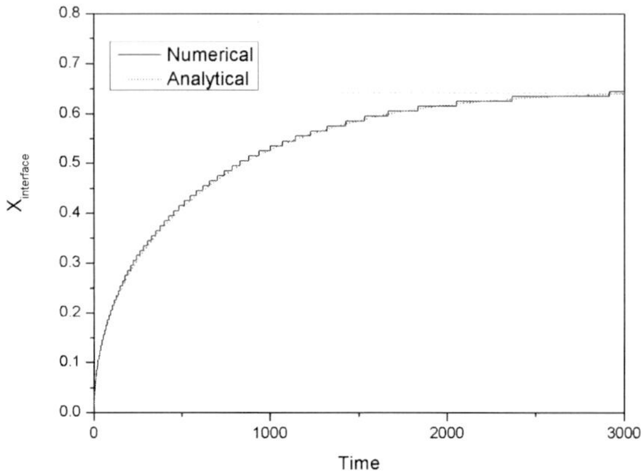

Figure 13. Validation of the numerical solution with the analytical solution.

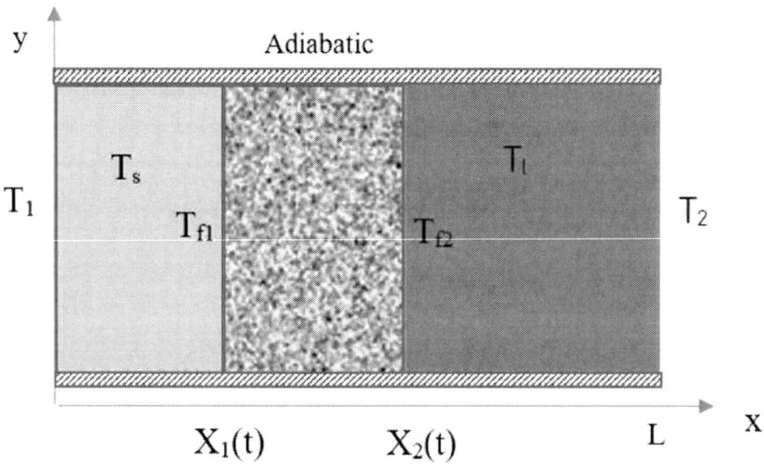

Figure 14. Three-phase problem in a rectangular cavity.

$$
\begin{aligned}
&X_2(t) = 2\lambda_2\sqrt{\alpha_2 t},\ for\ t\ \in [0, t_2] \\
&\frac{dX_2(t)}{dt} = \frac{\beta_3}{X_2(t) - X_1(t)} + \frac{\beta_4}{X_2(t) - L},\ for\ t \in [t_2, +\infty]
\end{aligned}
\tag{45}
$$

In which, β_1, β_2, β_3 and β_4 are given by the expressions:

$$\beta_1 = \frac{k_1(T_{f_1} - T_1)}{L_{H_1} \cdot \rho}, \beta_2 = \frac{k_2(T_{f_2} - T_{f_1})}{L_{H_1} \cdot \rho}$$
$$\beta_3 = \frac{k_2(T_{f_2} - T_{f_1})}{L_{H_2} \cdot \rho}, \beta_4 = \frac{k_3(T_2 - T_{f_2})}{L_{H_2} \cdot \rho}$$

And λ_1 and λ_2 are solutions of the equations below:

$$\frac{e^{-\lambda_1^2}}{erf\lambda_1} - \frac{k_2}{k_1}\sqrt{\frac{\alpha_1}{\alpha_2}}\frac{T_{f_2} - T_{f_1}}{T_{f_1} - T_1}\frac{e^{-\frac{\alpha_1}{\alpha_2}\lambda_1^2}}{erf\lambda_2 - erf(\lambda_1\sqrt{\frac{\alpha_1}{\alpha_2}})} = \frac{\lambda_1 L_{H_1}\sqrt{\pi}}{c_1(T_{f_1} - T_1)}$$
$$\frac{e^{-\lambda_2^2}}{erf\lambda_2 - erf(\lambda_1\sqrt{\frac{\alpha_1}{\alpha_2}})} - \frac{k_3}{k_2}\sqrt{\frac{\alpha_2}{\alpha_3}}\frac{T_2 - T_f}{T_{f_2} - T_{f_1}}\frac{e^{\frac{\alpha_2}{\alpha_3}\lambda_2^2}}{erf(\lambda_2\sqrt{\frac{\alpha_2}{\alpha_3}})} = \frac{\lambda_2 L_{H_2}\sqrt{\pi}}{c_2(T_{f_2} - T_{f_1})} \tag{46}$$

The connection times t_1 and t_2 are the solutions of the equations at the interfaces:

$$t_2 = \frac{L^2}{(2\lambda_2\sqrt{\alpha_2} - \frac{\beta_4}{\mu})^2}$$
$$\frac{\lambda_1\sqrt{\alpha_1}}{\sqrt{t_1}} = \frac{\beta_1}{2\lambda_1\sqrt{\alpha_1 t_1}} + \frac{\beta_2}{2\lambda_1\sqrt{\alpha_1 t_1} - X_2(t_1)} \tag{47}$$

Where μ is given by:

$$\mu = \lambda_2\sqrt{\alpha_2} + \frac{\alpha_3}{2(\lambda_1\sqrt{\alpha_1} - \lambda_2\sqrt{\alpha_2})} \tag{48}$$

With time going on, the interfaces will reach the asymptotic value:

$$X_{2\infty} = \frac{L}{1 + \frac{\beta_4}{\beta_3} \cdot \frac{\beta_2}{\beta_2 + \beta_1}} \tag{49}$$

Thus, this method of calculation is well adapted to the phase change problems of both the fluid and the porous case. It leads to the success in validating numerical simulations concerning the position of the phase change front with the analytical model inspired by Stefan's problem. The advantage it that it is able to provide results for a finite domain.

The three-phase case is only indicative because as indicated in the two-phase case, the validity of the semi-analytical solution must meet several criteria. The criterion is more important in the three-phase case. This is the major limitation in applying the latter solution.

3.6. Phase Change Problem in a Cylindrical Domain

A cylinder ($0 < r < R_0$) that contains liquid initially at a constant temperature T_1 ($T_1 > T_f$) is considered, and T_f is the melting temperature of the material [42]. From the moment $t = 0$, along the axis $r = 0$, Q as heat is extracted per unit time. This triggers a solidification process. The edge $r = R_0$ will remain at a temperature $T = T_1$.

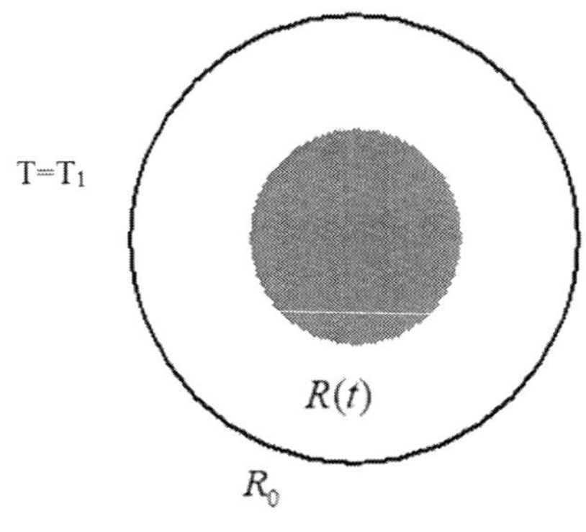

Figure 15. Gel problem with cylindrical symmetry: the source is placed on the line $r = 0$. The temperature at $r = R_0$ remains constant T_1.

$R(t)$ denotes the radius of the surface which separates the solid and liquid phases. Therefore, by using the analytical solution in the infinite domain given in Carslaw's research [7], one can easily apply the method described in the previous section to determine the position of this separation surface:

$$
\begin{aligned}
&R(t) = 2\lambda\sqrt{\alpha_1 t}, \; for \; t \in [0, t_0] \\
&\frac{dR(t)}{dt} = \frac{\gamma_1}{R(t)} + \frac{\gamma_2}{R(t)ln(R(t)/R_0)}, \; for \; t \in [t_0, +\infty]
\end{aligned} \tag{50}
$$

Where γ_1, γ_2 are given by:

$$
\gamma_1 = \frac{Q}{2\pi} L_H \rho, \; \gamma_2 = \frac{k_2(T_1 - T_f)}{L_H \rho} \tag{51}
$$

The connection time t_0 is given by:

$$
t_0 = \frac{R_0^2}{2\eta_0} exp(-2\frac{\gamma_2}{\gamma_1} - \eta_0) \tag{52}
$$

The asymptotic value of $R(t)$ is: $R_\infty = R_0 exp(-\frac{\gamma_2}{\gamma_1})$.

Conclusion

In this chapter, the energy storage properties of the phase change materials are introduced. PCMs provide an approach to improve the energy efficiency and thermal comfort in buildings. The fundamental properties and applications of PCMs are presented, as well as the existing solutions for phase change problems. Some analytical solutions are recalled afterwards, including Stefan problem in 1D semi-infinite domain, the case of 2D rectangular corner, a global energy estimation in a finite domain for freezing process and a semi-analytical solution in bounded domain. The semi-analytical solution is carried out based on the connection method between solutions at the short time and the stationary case in order to achieve the phase change mechanism in a finite domain. Finally, the the multi-phase problem and phase change process in a cylindrical domain are presented as well.

References

[1] Liu Z., Yu Z., Yang T., Qin D., Li S., Zhang G., Haghighat F. and Joybari M. M., A review on macro-encapsulated phase change material for building envelope applications, *Build. Environ.* **144**, 281–294 (2018).

[2] Delgado M., Lázaro A., Biedenbach M., Gamisch S., Gschwander S., Höhlein S., Haagen A. K. and Bríggemann D., Intercomparative tests on viscosity measurements of phase change materials, *Thermochim. Acta* **668**, 158–168 (2018).

[3] Rao V. V., Parameshwaran R. and Ram V. V., PCM-mortar based construction materials for energy efficient buildings: A review on research trends, *Energy Build.* **158**, 95–122 (2018).

[4] Song M., Niu F, Mao N., Hu Y. and Deng S., Review on building energy performance improvement using phase change materials, *Energy Build.* **158**, 776–793 (2018).

[5] Zeinelabdein R., Omer S. and Gan G., Critical review of latent heat storage systems for free cooling in buildings, *Renew. Sustain. Energy Rev.* **82**, 2843–2868 (2018).

[6] Farid M., Khudhair A. M., Razack S. A. and Hallaj S. A., A review on phase change energy storage: materials and applications, *Energy Convers. Manage.* **45**, 1597-1615 (2004).

[7] Carslaw H. S., Jaeger J. C., Conduction of heat in solids, *Oxford, Clarendon Press*, 282-286 (1959).

[8] Barreneche C., Navarro M. E., Fernandez A. I. and Cabeza L. F., Improvement of the thermal inertia of building materials incorporating PCM. Evaluation in the macroscale, *Appl. Energy* **109**, 428–432 (2013).

[9] Evola G., Marletta L. and Sicurella F., A methodology for investigating the effectiveness of PCM wallboards for summer thermal comfort in buildings, *Build. Environ.* **59**, 517–527 (2013).

[10] Farid M. M., Sherrif A., 16 - Phase change materials for energy storage and thermal comfort in buildings, Materials for Energy Efficiency and Thermal Comfort in Buildings, *Woodhead Publishing Series in Energy*, United Kingdom, 384–398 (2010).

[11] Medina M. A, King J. B. and Zhang M., On the heat transfer rate reduction of structural insulated panels (SIPs) outfitted with phase change materials (PCMs), *Energy*, Elsevier **33**, 667–678 (2008).

[12] Raud R., Jacob R., Bruno F., Will G. and Steinberg T. A., A critical review of eutectic salt property prediction for latent heat energy storage systems, *Renew. Sustain. Energy Rev.* **70**, 936–944 (2017).

[13] Cunha J. P., Eames P., Thermal energy storage for low and medium temperature applications using phase change materials: A review, *Appl. Energy* **177**, 227–238 (2016).

[14] Wang X., Yap C. and Mujumdar A. S., A parametric study of phase change material (PCM)-based heat sinks, *Int. J. Therm. Sci.* **16**, 1055-1068 (2008).

[15] Royon L, Karim L. and Bontemps A., Thermal energy storage and release of a new component with PCM for integration in floors for thermal management of buildings, *Energy Build.* **63**, 29-35 (2013).

[16] Zhou D., Zhao C. Y. and Tian Y., Review on thermal energy storage with phase change materials (PCMs) in building applications, *Appl. Energy* **92**, 593-605 (2012).

[17] Laaouatni A., Martaj N., Bennacer R., Omari M. E. and Ganaoui M. E., Phase change materials for improving the building thermal inertia, *Energy Procedia* **139**, 744–749 (2017).

[18] Memon S. A., Phase change materials integrated in building walls: a state of the art review, *Renew. Sustain. Energy Rev.* **31**, 870–906 (2014).

[19] Saikia P., Azad A. S. and Rakshit D., Thermodynamic analysis of directionally influenced phase change material embedded building walls, *Int. J. Therm. Sci.* **126**, 105–117 (2018).

[20] Peippo K., Kauranen P. and Lund P. D., A multicomponent PCM wall optimized for passive solar heating, *Energy Build.* **17**, 259–270 (1991).

[21] Khudhair A. M., Farid M. M., A review on energy conservation in building applications with thermal storage by latent heat using phase change materials, *Energy Convers. Manage.* **45**, 263-275 (2004).

[22] Royon L., Karim L. and Bontemps A., Optimization of PCM embedded in a floor panel developed for thermal management of the lightweight envelope of buildings, *Energy build.* **82**, 385–390 (2014).

[23] Lin K., Zhang Y., Xu X., Di H., Yang R. and Qin P., Modeling and simulation of under-floor electric heating system with shape-stabilized PCM plates, *Build. Environ.* **39**, 1427-1434 (2004).

[24] Castell A., Martorell I., Medrano M., Pérez G. and Cabeza L. F., Experimental study of using PCM in brick constructive solutions for passive cooling, *Energy Build.* **42**, 534–540 (2010).

[25] Principi P., Fioretti R., Thermal analysis of the application of PCM and low emissivity coating in hollow bricks, *Energy Build.* **51**, 131–142 (2012).

[26] Pasupathy A., Athanasius L., Velraj R. and Seeniraj R. V., Experimental investigation and numerical simulation analysis on the thermal performance of a building roof incorporating phase change material (PCM) for thermal management, *Appl. Therm. Eng.* **28**, 556–565 (2008).

[27] Goia F., Perino M., Serra V., Improving thermal comfort conditions by means of PCM glazing systems, *Energy Build.* **60**, 442–452 (2013).

[28] Nagano K., Takeda S., Mochida T., Shimakura K., Nakamura T., Study of a floor supply air conditioning system using granular phase change material to augment building mass thermal storage: Heat response in small scale experiments, *Energy Build.* **38**, 434–446 (2006).

[29] Waqas A., Din Z. U., Phase change material (PCM) storage for free cooling of buildings - A review, *Renew. Sustain. Energy Rev.* **18**, 607–625 (2013).

[30] Ahmad M., Bontemps A., Sallé H. and Quenard D., Experimental investigation and computer simulation of thermal behaviour of wallboards containing a phase change material, *Energy Build.* **38**, 357–366 (2006).

[31] Zhang D., Li Z., Zhou J. and Wu K., Development of thermal energy storage concrete, *Cem. Concr. Res.* **34**, 927–934 (2004).

[32] Bart J. C., Hoogendoorn C. J. and Schaareman P. B., A characteristic dimensionless time in phase change problems, *J. Sol. Energy Eng.* **108**, 310–315 (1986).

[33] Budhia H., Kreith F., Heat transfer with melting or freezing in a wedge, *Int. J. Heat Mass Transfer* **16**, 195–211 (1973).

[34] Rathjen K. A., Jiji L. M., Heat conduction with melting or freezing in a corner, *J. Heat Transfer* **93**, 101–109 (1971).

[35] Budhia H., Kreith F., Heat transfer with melting or freezing in a wedge, *Int. J. Heat Mass Transfer* **16**, 195–211 (1973).

[36] Bart G. C. J., Estimation of freezing or chilling behavior, *Int. J. Refrigeration* **21**, 55–63 (1998).

[37] Benzadi M., *Contribution l'étude numéique et expérimentale de changement de phase en milieu poreux saturé: Généralisation de la formulation enthalpique* [*Contribution to the numerical and experimental study of phase change in saturated porous media: Generalization of the enthalpic formulation*], PhD, Cergy Pontoise University, France (1999).

[38] Kim C. J., Kaviany M., A numerical method for phase-change problems, *Int. J. Heat Mass Transfer* **33**, 2721–2734 (1990).

[39] Habibullin I. T., Symmetry approach in boundary value problems, *J. Nonlinear Math. Phys.* **3**, 147–151 (1996).

[40] Ogoh W, Groulx D., Stefan's problem: Validation of a one-dimensional solid-liquid phase change heat transfer process, *Proceeding Comsol Conference* (2010).

[41] Qarnia M. E., Adnani F. E. and Lakhal E. K., Approximate analytical solution for one-dimensional solidification problem of a finite superheating phase change material including the effects of wall and thermal contact resistance, *J. Appl. Math.*, 1–20 (2012).

[42] Boureghda A., Solution of an iced melting cylindrical problem, *J. Nonlinear Sci. Appl.* **9**, 1440–1452 (2015).

About the Editors

Jordan Hristov, PhD, DSc
Professor of Chemical Engineering
Department of Chemical Engineering
University of Chemical Technology and Metallurgy,
Sofia, Bulgaria
Email: jordan.hristov@mail.bg; hristovmeister@gmail.com

Rachid Bennacer, PhD
Professor
LMT-ENS Cachan, CNRS
Paris-Saclay University
Cachan, France

INDEX

A

B

C

D

E

F

G

H

I

L

M

N

T

V

W

Y

Metals and Metal-Based Electrocatalytic Materials for Alternative Energy Sources and Electronics

Editors: Jasmina Stevanović, Ph.D. (Institute of Chemistry, Technology and Metallurgy, University of Belgrade, Belgrade, Serbia)

Series: Chemical Engineering Methods and Technology

Book Description: The important role of metals, their oxides and catalytically-interactive supports in contemporary investigations related to rational construction of next-generation devices as alternative energy sources and hi-tech electronics is ambitiously presented throughout this book.

Hardcover ISBN: 978-1-53614-663-9
Retail Price: $230

Centrifugal Pumps: Characteristics, Uses and Performance

Editors: Ward Vaughn and Jonathan Hector

Series: Chemical Engineering Methods and Technology

Book Description: In Chapter One, Fujun Wang, Chaoyue Wang, Zhengjun Yang, Peijian Zhou, and Zhifeng Yao propose and examine a dynamic mixed nonlinear SGS model, with the results indicating that this wall-resolved near-wall solution could capture details more accurately.

Hardcover ISBN: 978-1-53612-546-7
Retail Price: $230